OUT OF THIS WORLD

Developed and Published by

AIMS Education Foundation

© 2007 AIMS Education Foundation

Education Foundation

This book contains materials developed by the AIMS Education Foundation. **AIMS** (**A**ctivities **I**ntegrating **M**athematics and **S**cience) began in 1981 with a grant from the National Science Foundation. The non-profit AIMS Education Foundation publishes hands-on instructional materials that build conceptual understanding. The foundation also sponsors a national program of professional development through which educators may gain expertise in teaching math and science.

AIMS Education Foundation
P.O. Box 8120, Fresno, CA 93747-8120
888.733.2467 • aimsedu.org

ISBN 978-1-932093-13-1

Printed in the United States of America

Table of Contents

Table of Contents

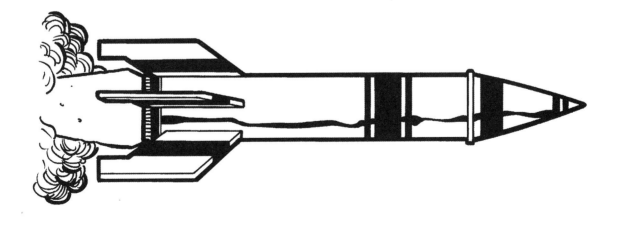

I Hear
and
I Forget,

I See
and
I Remember,

I Do
and
I Understand.

Chinese Proverb

Project 2061 Benchmarks

The Nature of Mathematics
- *Mathematical ideas can be represented concretely, graphically, and symbolically.*
- *Numbers and shapes—and operations on them—help to describe and predict things about the world around us.*

The Physical Setting
- *The earth is one of several planets that orbit the sun, and the moon orbits around the earth.*
- *Stars are like the sun, some being smaller and some larger, but so far away that they look like points of light.*
- *Things on or near the earth are pulled toward it by the earth's gravity.*
- *Like all planets and stars, the earth is approximately spherical in shape. The rotation of the earth on its axis every 24 hours produces the night-and-day cycle. To people on earth, this turning of the planet makes it seem as though the sun, moon, planets, and stars are orbiting the earth once a day.*
- *Nine planets of very different size, composition, and surface features move around the sun in nearly circular orbits. Some planets have a great variety of moons and even flat rings of rock and ice particles orbiting around them. Some of these planets and moons show evidence of geological activity. The earth is orbited by one moon, many artificial satellites, and debris.*
- *We live on a relatively small planet, the third from the sun in the only system of planets definitely known to exist (although other, similar systems may be discovered in the universe).*
- *The moon's orbit around the earth once in about 28 days changes what part of the moon is lighted by the sun and how much of that part can be seen from the earth—the phases of the moon.*
- *The moon looks a little different every day, but looks the same again about every four weeks.*
- *The sun can be seen only in the daytime, but the moon can be seen sometimes at night and sometimes during the day. The sun, moon, and stars all appear to move slowly across the sky.*

The Designed World
- *Communication technologies make it possible to send and receive information more and more reliably, quickly, and cheaply over long distances.*

The Mathematical World
- *Tables and graphs can show how values of one quantity are related to values of another.*
- *Graphical display of numbers may make it possible to spot patterns that are not otherwise obvious, such as comparative size and trends.*
- *The graphic display of numbers may help to show patterns such as trends, varying rates of change, gaps, or clusters.*

Such patterns sometimes can be used to make predictions about the phenomena being graphed.
- *Find the mean and median of a set of data.*
- *The mean, median, and mode tell different things about the middle of a data set.*
- *The larger a well-chosen sample is, the more accurately it is likely to represent the whole. but there are many ways of choosing a sample that can make it unrepresentative of the whole.*
- *Mathematical statements can be used to describe how one quantity changes when another changes. Rates of change can be computed from magnitudes and vice versa.*
- *Scale drawings show shapes and compare locations of things very different in size.*

Common Themes
- *In something that consists of many parts, the parts usually influence one another.*
- *A system can include processes as well as things.*
- *Geometric figures, number sequences, graphs, diagrams, sketches, number lines, maps, and stories can be used to represent objects, events, and processes in the real world, although such representations can never be exact in every detail.*
- *Things change in steady, repetitive, or irregular ways—or sometimes in more than one way at the same time. Often the best way to tell which kinds of change are happening is to make a table or graph of measurements.*
- *Different models can be used to represent the same thing. What kind of a model to use and how complex it should be depends on its purpose. The usefulness of a model may be limited if it is too simple or if it is needlessly complicated. Choosing a useful model is one of the instances in which intuition and creativity come into play in science, mathematics, and engineering.*
- *Seeing how a model works after changes are mode to it may suggest how the real thing would work if the same were done to it.*
- *Things that change in cycles, such as the seasons or body temperature, can be described by their cycle length or frequency, what the highest and lowest values are, and when they occur. Different cycles range from many thousands of years down to less than a billionth of a second.*

Habits of Mind
- *Add, subtract, multiply, and divide whole numbers mentally, on paper, and with a calculator.*
- *Make sketches to aid in explaining procedures or ideas.*
- *Use calculators to compare amounts proportionally.*
- *Organize information in simple tables and graphs and identify relationships they reveal.*
- *Use numerical data in describing and comparing objects and events.*

Benchmarks for Science Literacy: Project 2061
American Association for the Advancement of Science
Oxford University Press
New York. 1993.

NRC Standards

Abilities Necessary to do Scientific Inquiry
- *Develop descriptions, explanations, predictions, and models using evidence.*
- *Use appropriate tools and techniques to gather, analyze, and interpret data.*
- *Use mathematics in all aspects of scientific inquiry.*

Understandings About Scientific Inquiry
- *Different kinds of questions suggest different kinds of scientific investigations. Some investigations involve observing and describing objects, organisms, or events; some involve collecting specimens; some involve experiments; some involve seeking more information; some involve discovery of new objects and phenomena; and some involve making models.*
- *Mathematics is important in describing and comparing objects and events.*

Position and Motion of Objects
- *The position of an object can be described by locating it relative to another object or the background.*

Objects in the Sky
- *The sun, moon, stars, clouds, birds, and airplanes all have properties, locations, and movements that can be observed and described.*

Earth in the Solar System
- *The earth is the third planet from the sun in a system that includes the moon, the sun, eight other planets and their moons, and smaller objects, such as asteroids and comets. The sun, an average star, is the central and largest body in the solar system.*
- *Most objects in the solar system are in regular and predictable motion. Those motions explain such phenomena as the day, the year, phases of the moon, and eclipses.*
- *Gravity is the force that keeps planets in orbit around the sun and governs the rest of the motion in the solar system. Gravity alone holds us to the earth's surface and explains the phenomena of the tides.*

National Science Education Standards
National Research Council
National Academy Press
Washington, D.C. 1996.

NCTM Standards 2000*

Number and Operations
- *Select appropriate methods and tools for computing with whole numbers from among mental computation, estimation, calculators, and paper and pencil according to the context and nature of the computation and use the selected method or tools*
- *Understand and use ratios and proportions to represent quantitative relationships*

Algebra
- *Model problem situations with objects and use representations such as graphs, tables, and equations to draw conclusions*

Geometry
- *Identify, compare, and analyze attributes of two- and three-dimensional shapes and develop vocabulary to describe the attributes*
- *Make and test conjectures about geometric properties and relationships and develop logical arguments to justify conclusions*
- *Describe location and movement using common language and geometric vocabulary*
- *Build and draw geometric objects*
- *Use geometric models to solve problems in other areas of mathematics, such as number and measurement*
- *Recognize geometric ideas and relationships and apply them to other disciplines and to problems that arise in the classroom or in everyday life*

Measurement
- *Recognize the attributes of length, volume, weight, area, and time*
- *Understand such attributes as length, area, weight, volume, and size of angle and select the appropriate type of unit for measuring each attribute*
- *Carry out simple unit conversions, such as from centimeters to meters, within a system of measurement*
- *Understand that measurements are approximations and how differences in units affect precision*
- *Select and apply appropriate standard units and tools to measure length, area, volume, weight, time, temperature, and the size of angles*
- *Solve problems involving scale factors, using ratio and proportion*
- *Solve simple problems involving rates and derived measurements for such attributes as velocity and density*

Data Analysis and Probability
- *Collect data using observations, surveys, and experiments*
- *Represent data using tables and graphs such as line plots, bar graphs, and line graphs*
- *Use measures of center, focusing on the median, and understand what each does and does not indicate about the data set*
- *Propose and justify conclusions and predictions that are based on data and design studies to further investigate the conclusions or predictions*

Problem Solving
- *Solve problems that arise in mathematics and in other contexts*

Communication
- *Communicate their mathematical thinking coherently and clearly to peers, teachers, and others*

Representation
- *Use representations to model and interpret physical, social, and mathematical phenomena*

* Reprinted with permission from *Principles and Standards for School Mathematics*, 2000 by the National Council of Teachers of Mathematics. All rights reserved.

Principles and Standards for School Mathematics
National Council of Teachers of Mathematics
Reston, VA. 2000.

Rubber Band Books: Reading in the Content Area

The Rubber Band Books offer valuable content information presented in a kid-friendly way. Each student can be given his or her own book to keep and refer to at a later date. These books also provide a great home link, as students can take them home and share the information they are learning with their parents. To assemble a book, follow these simple instructions:

Fold back

Then over

Nest together

Hold together with a large rubber band

A #19 rubber band fits perfectly. If these are not available, clip off the inside corners of the book so that the other rubber bands can fit.

Then there was Sir Isaac Newton, an English mathematician and physicist. In 1686, Newton published three laws that related forces, matter, and motion and from these he came up with the law of gravity. Other scientists used his law of gravity in their study of astronomy.

Way, way, way back in time, around 100 A.D., Ptolemy, an astronomer in Alexandria, Egypt, made an organized chart of how the known planets and sun orbited the Earth. Ptolemy lived before the telescope was invented, so all observations at his time were made with the naked eye.

Throughout time, more great thinkers have added to our knowledge about what we know about the universe. Among them are scientists like Edmund Halley, Albert Einstein, and Stephen Hawking. With increasing technology and scientific thinking, we will continue to revise our ideas about this cosmos in which we live.

Einstein

Hawking

Halley

A VERY BRIEF HISTORY OF ASTRONOMY

OUT OF THIS WORLD

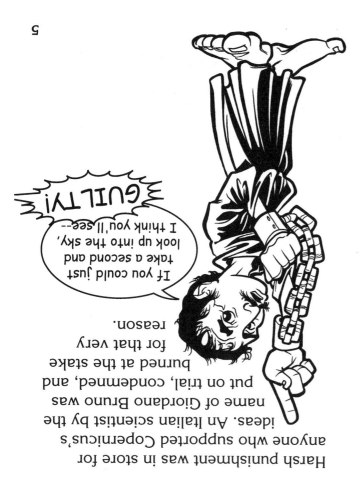

GUILTY!

If you could just take a second and look up into the sky, I think you'll see-- reason.

burned at the stake for that very reason.

Harsh punishment was in store for anyone who supported Copernicus's ideas. An Italian scientist by the name of Giordano Bruno was put on trial, condemned, and burned at the stake for that very reason.

the center (of the solar system).

* Hello is taken from the Greek word meaning sun; therefore, heliocentric means the sun is the center (of the solar system).

* This reasoning is known as heliocentric.

astronomy, the study of the universe.

This was a very shocking new theory. It went against the beliefs of so many people. In honor of his work, he is considered the founder of modern astronomy, the study of the universe.

COPERNICUS

In the 1500s, Copernicus wrote that the Earth rotated on its axis once daily and traveled around the sun once yearly.*

This geocentric way of thinking remained for over a thousand years.

Another Italian by the name of Galileo believed the heliocentric theory. With a spyglass that he made, he saw mountains on the moon, that the Milky Way was made of stars, and that there were objects orbiting Jupiter. Sometime around 1630, Galileo was warned that he should not defend Copernicus's theory and was sentenced to house arrest for the rest of his life.

Ptolemy's system was used to predict the position of the planets that he thought orbited the Earth* in the order Mercury, Venus, Sun, Mars, Jupiter, and Saturn.

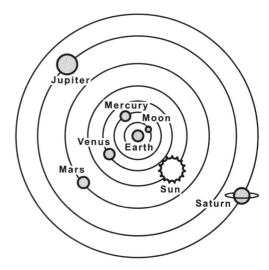

* This reasoning was called a geocentric theory of the solar system. Geo is a prefix meaning Earth, centric means the center of; thus, geocentric means the Earth is the center (of the solar system).

HOW FAR TO THE SUN?

0-7041-776

Topic
Size of and distance to the sun

Key Question
How far is it to the Sun?

Learning Goals
Students will:
- build scale models of the Sun and Earth, and
- place the Sun and Earth the approximate scaled distance apart.

Guiding Documents
Project 2061 Benchmarks
- *Mathematical ideas can be represented concretely, graphically, and symbolically.*
- *The earth is one of several planets that orbit the sun, and the moon orbits around the earth.*
- *Scale drawings show shapes and compare locations of things very different in size.*
- *Use numerical data in describing and comparing objects and events.*

NRC Standard
- *The earth is the third planet from the sun in a system that includes the moon, the sun, eight other planets and their moons, and smaller objects, such as asteroids and comets. The sun, an average star, is the central and largest body in the solar system.*

NCTM Standards 2000*
- *Use geometric models to solve problems in other areas of mathematics, such as number and measurement*
- *Understand such attributes as length, area, weight, volume, and size of angle and select the appropriate type of unit for measuring each attribute*

Math
Measuring
    diameter
    length
Proportionality

Science
Earth science
    solar system

Materials
Butcher paper
Chalk
String
Meter tape
Scissors

Background Information
Students often have questions about the sun's size and the heat it generates. They have "learned" that the sun is very large, that it is a source of energy on Earth, and that it is very hot. They often fail to realize that because of the great distance between the Earth and the sun that the Earth receives only a very small portion (about one two-billionth of the total energy released by the sun and that this great distance also makes the sun appear to be much smaller than it actually is. In this activity, it is hoped that students will better be able to conceptualize the relative sizes of and distance between the sun and Earth.

Management
1. Students should work in groups.
2. Be sure students know how to use string loops and pencils for drawing large circles.
3. In order to space out the scaled distance between the sun and Earth, you will need to find an area that is at least 55 meters in length. You may want to measure this distance as a whole class project rather than as a small group project.

Procedure
1. Ask students what they know about the size of the sun. Ask them what they know about the temperature of the sun.
2. Ask students why the sun is important to us on Earth.
3. Explain to the students that the sun is much bigger than our Earth. It is about 109 times bigger. Have students explain why it seems so small if it really is so much larger. [It is a long distance away.]
4. Tell the students that they are going to make scale models of the Earth and the Sun to compare sizes. They will then try to predict how far apart they would be according to the same scale.
5. Direct them to cut the Earth and sun models as indicated on their activity sheet.

6. Once the models are cut, ask the students that if the sun is so hot and so big, why we don't burn up? [The sun is about 93 million miles from the Earth, or about 100 sun models according to our scale. (55 cm x 100 = 5500cm or 55 meters.)]
7. Have the students predict and record the distance the sun is from the Earth.
8. Take the students to the designated area and measure the 55 meters that represents the scaled distance between the sun and Earth.

Connecting Learning
1. What did you learn about the sizes of the sun and the Earth?
2. Why don't we burn up from the heat of the sun?
3. What role does the sun play in life on Earth? [light, heat, energy for the growth of green plants, major role in the food chain, helps drive the water cycle, etc.]
4. What are you wondering now?

Extension
To help students better understand the sun's energy, light a candle in a darkened room. Ask students what they observe about the light from the candle. [It goes in all directions.] Apply this to the sun and how its light spreads in all directions; its light energy doesn't all go directly to the Earth. Have students hold their hands near the flame of the candle to see that there is both heat and light energy coming from it which is similar to the sun.

* Reprinted with permission from *Principles and Standards for School Mathematics*, 2000 by the National Council of Teachers of Mathematics. All rights reserved.

0033

OUT OF THIS WORLD

001001010110100100101010
HOW FAR TO THE SUN?
0010010101101001001010
0-7041-776

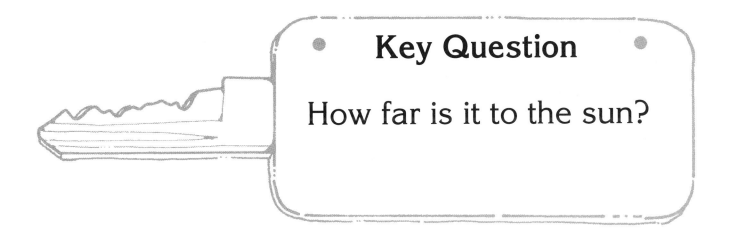

Key Question

How far is it to the sun?

Learning Goals

Students will:

- build a scale model of the sun and Earth, and

- place them the approximate scaled distance apart.

HOW FAR TO THE SUN?

YOU WILL NEED:

9 regular paper clips

meter tape

paper

scissors

2 pencils

DO THIS:

1. Make a paper model of the sun and Earth.

   Earth—Cut a paper circle this big. (.5 cm) Earth

   Sun—Cut a paper circle 55 cm in diameter. Use 2 pencils and 9 paper clips to draw it.

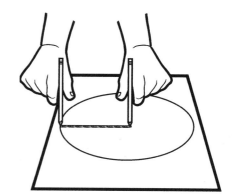

2. Using the same scale as the models, estimate how far the sun is from the Earth. Place the two models that far apart.

   Measure the distance between.

   My guess:_____

3. The Earth is 93,000,000 miles from the sun. Start at the Earth and walk the model of the sun 100 suns away.

   Measure the distance between.

   My measurement:_____

4. Glue your Earth to the sun. Wow!

5. If the sun is so much larger than Earth, why does it look so small?

Connecting Learning

1. What did you learn about the sizes of the sun and Earth?

2. Why don't we burn up from the heat from the sun?

3. What role does the sun play in our daily lives?

4. How do models help us learn about things?

5. What are you wondering now?

APPARENT SIZES

Topic
Apparent sizes of the sun and moon

Key Question
How can you make four different-sized objects appear to be the same size?

Learning Goals
Students will:

- explore why the sun and the moon appear to be about the same size in the sky, and
- identify how distance influences the apparent size of objects in the sky.

Guiding Documents
Project 2061 Benchmarks

- *Stars are like the sun, some being smaller and some larger, but so far away that they look like points of light.*
- *Seeing how a model works after changes are made to it may suggest how the real thing would work if the same were done to it.*

NRC Standards

- *The position of an object can be described by locating it relative to another object or the background.*
- *The sun, moon, stars, clouds, birds, and airplanes all have properties, locations, and movements that can be observed and described.*

NCTM Standards 2000*

- *Model problem situations with objects and use representations such as graphs, tables, and equations to draw conclusions*
- *Represent data using tables and graphs such as line plots, bar graphs, and line graphs*

Math
Measurement
Relative sizes

Science
Earth science
   astronomy
      size and distance relationship

Integrated Processes
Observing
Comparing and contrasting
Inferring
Predicting
Communicating

Materials
Tubes (empty toilet paper tubes work best)
Four round objects of different sizes (for example: marble, table tennis ball, tennis ball, volleyball)
Black paper
Paper punch

Background Information
Two different-sized objects may *appear* to be the same size or two same-sized objects may *appear* to be different sizes. This appearance of the objects' size is caused by the objects being different distances from the viewer. The farther something is from the viewer, the smaller it will appear.

From Earth, the *apparent sizes* of the sun and the moon seem to be approximately the same. We "know" that there is a vast difference in the sizes and distances from the Earth of both objects. The moon is a body only 3476 km in diameter, while the sun is approximately 400 times that diameter—1,400,000 km. But the sun is also 400 times as far away as the moon. The distance of the sun from the Earth is approximately 150,000,000 km and the distance of the moon from the Earth is 384,000 km.

A small object can cover up a larger object if the small object is close enough to the eye. A penny held at arm's length is big enough to cover the disk of the sun and also the disk of the moon. So, although the moon is much smaller than the sun, it is close enough to the Earth to appear the same size.

You can help students perceive how size and distance can affect how things look by challenging them to look through a hole in a tube at different-sized spheres and move their position until the spheres appear to be the same size as the paper hole.

Management

1. Obtain four spherical objects ranging in size from small to large. The balls mentioned in the *Materials* section are only suggestions.
2. Cut a piece of black construction paper large enough to cover one end of the tube. Using a paper punch, put a hole in the middle of the paper. Secure the paper to the tube using tape or a rubber band. Make certain the hole is centered on the end of the tube.
3. To properly use the tube to look at the objects, tell the students to put the open end to their eye, resting it on their cheekbone.

Procedure

1. Ask students how many of them have seen a full moon. Discuss with them how the size of the full moon compares to the size of the sun.
2. Place the four spherical objects on a flat surface. Demonstrate the proper way to use the tubes. Ask students to predict whether or not they think they will be able to see the objects (only the small? ...only the large? ...all four? etc.) through the hole. After predictions and discussion, direct them to look at the objects through the tube.
3. After some exploration time, ask them where they think they will have to stand to make one of the objects appear to be the same size as the hole. Encourage them to consider whether they will be able to see all four objects by standing in that one position.
4. Let the students walk backward or forward as needed until the objects just fill the small hole in the paper. Ask them to give their reactions to what they found out. Emphasize the word *appears*. For example: From this position, the volley ball *appears* to be the same size as the marble *appears* from this other position.
5. Repeat the activity, but this time have the viewing students stand still and have others hold the objects. Tell the viewing students to give instructions to the object holders as to where they need to position themselves to make the object appear to fill the hole of the tube.
6. Ask the students to relate what they found out to their knowledge of the size and distance of the sun and moon. If necessary, select two spheres (one large and one small) to represent the sun and the moon respectively. Ask students to determine how far away the sun must be to fill the hole in the tube and how far away the moon must be to fill the hole in the tube.
7. Have the students record their findings on the activity page or in their science journal.

Connecting Learning

1. What do we mean by apparent size? [The size something appears to be.]
2. Is apparent size the same as actual size? Explain.
3. How does distance make apparent size smaller or larger?
4. How does this activity explain the fact that the sun and the moon appear to be about the same size?
5. Which one of the objects do you think might represent the sun? Explain.
6. What are you wondering now?

Extension

Try the same activity using the longer tubes such as those from rolls of paper towel. Have students explore whether the length of the viewing tube makes any difference in what they see and where they stand. Ask for explanations.

* Reprinted with permission from *Principles and Standards for School Mathematics,* 2000 by the National Council of Teachers of Mathematics. All rights reserved.

Key Question

How can you make four different-sized objects appear to be the same size?

Learning Goals

Students will:

- explore why the sun and the moon appear to be about the same size in the sky, and

- identify how distance influences the apparent size of objects in the sky.

APPARENT SIZES

05-182-005

▼ Describe what you had to do to see the objects.

▼ Explain why you think you had to do it.

▼ Explain how this activity relates to the apparent sizes of the sun and moon.

0021

0010010101101001001010

OUT OF THIS WORLD

APPARENT SIZES

0010010101101001001010

05-182-005

Connecting Learning

1. What do we mean by apparent size?

2. Is apparent size the same as actual size?

3. In your own words, explain why the sun and the moon appear to be the same size?

4. How does distance make apparent size smaller or larger?

5. Which one of the objects do you think might represent the sun? Explain.

6. What are you wondering now?

DIZZY SPELLS

Topic
Earth's rotation

Key Question
Why do we have daylight and nighttime?

Learning Goal
Students will discover how the Earth's rotation causes daylight and nighttime.

Guiding Documents
Project 2061 Benchmark
- *Like all planets and stars, the earth is approximately spherical in shape. The rotation of the earth on its axis every 24 hours produces the night-and-day cycle. To people on earth, this turning of the planet makes it seem as though the sun, moon, planets, and stars are orbiting the earth once a day.*

NRC Standard
- *Most objects in the solar system are in regular and predictable motion. These motions explain such phenomena as the day, the year, phases of the moon, and eclipses.*

Science
Earth science
 rotation
  daylight and night
Safety

Integrated Processes
Observing
Comparing and contrasting
Applying

Materials
Swivel chair or stool
Light source or brightly colored paper sun
Sun Shades (see *Management 2*)

Background Information
 Nearly every 24 hours, the Earth makes a complete rotation on its axis. It is this rotation that gives us daylight and nighttime. Simply put, as we turn, we are in varying amounts of the light from the sun. At dawn, the location we are in rotates into the path of the sun's rays and the light is dim. As the Earth continues its rotation, the light gains in intensity. When we more directly face the sun (midday), the light is bright. As the rotation continues, we begin to move out of these rays (dusk) and the light becomes dim. At midnight, for most of us, we are in the shadow of the Earth and it is dark. Children may think there is no light, but the darkness is caused because the Earth is blocking the sunlight. It is an opaque object that blocks the lights rays.

 It is difficult to believe that we are on a moving planet. The Earth rotates at an amazing speed, approximately 1670 km/hr (1070 mi/hr) at the equator. At 45 degrees latitude, the speed is 1180 km/hr (756 mi/hr). Children often wonder what would happen if the Earth stopped spinning. If it stopped spinning completely, we would have half a year of daylight and half a year of nighttime. If it stopped spinning suddenly, the atmosphere would keep spinning at its present rate and the winds would strip the Earth of all loose surface coverings. The Earth would also lose some of its magnetic field values. These snippets are not important to this lesson; however, students may appreciate the awe and wonder of something that we take for granted and don't realize is happening.

Management
1. This is a simple activity to demonstrate why we have daylight and darkness. Make certain that the students understand the safety factors involved in spinning around on the swivel chair or stool.
2. Run the *Sun Shades* on card stock.
3. Either have a bright light source or hang a large paper circle on the wall to represent the sun.

Procedure
1. Ask the *Key Question* and state the *Learning Goal.*
2. Ask the students to describe day. Make certain they state that the sun is our source of light. Tell them that they will be looking at the Earth/sun relationship to understand daylight and nighttime.
3. Point out the sun. Inform them that when they can see the sun, it will be daylight and when they can't see it, it will be nighttime.
4. Invite everyone to stand so that it is daylight.
5. Invite them to stand so that it is night.
6. Ask them if the sun is still out at night.

7. Arrange the swivel chair or stool so that a student sitting on it can see the sun. Ask that student to remain seated but to rotate his/her position so that the sun cannot be seen.
8. Point out to students that the observer had to spin the chair around. Inform them that the Earth spins much like this. Ask the observer to hold the *Sun Shades* next to his/her face while slowly spinning around in the chair.
9. Continually ask the observer how much of the sun he or she is seeing. Relate to students that it is the same for us on Earth, sometimes we're in the light of the sun and sometimes we're not.

10. Have students rotate on the chair, moving to the left. Discuss dawn, midday, dusk, and night. Let all students have a turn doing this.
11. Distribute the student page and ask students to describe what time of day is represented in the illustrations.

Connecting Learning
1. What causes day and night?
2. What is dawn?
3. What is dusk?
4. Why is it dark at midnight (in most places)?
5. Why is it brightest at midday?
6. What does the word *rotation* mean?
7. What are you wondering now?

DIZZY SPELLS

Key Question

Why do we have daylight and nighttime?

Learning Goal

Students will:

discover how the Earth's rotation causes daylight and nighttime.

15

DIZZY SPELLS

CHOOSE FROM THESE:

!Dawn !Dusk !Midday !Night

Time of Day

Time of Day

Time of Day

Time of Day

_____   _____   _____   _____

16

DIZZY SPELLS

Sun Shades

Connecting Learning

1. What causes day and night?

2. What is dawn?

3. What is dusk?

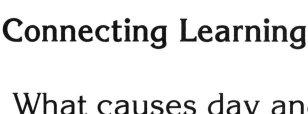

4. Why is it dark at midnight (in most places)?

5. Why is it brightest at midday?

6. What does the word *rotation* mean?

7. What are you wondering now?

The sun basically stays in place and the Earth rotates. (The Earth also revolves around the sun, but that's another story.) Each rotation gives us day and night. You can use that tennis ball model to understand why people in New York see the sun rise before people in San Francisco.

Does the sun shine in your bedroom window in the morning? Is there a favorite place where you like to sit in the sun during the winter? When you walk to school, where is the sun? When you walk home from school, where is the sun?

When we talk about the sun moving across the sky, we refer to it as apparent motion. Apparent motion means that it appears that way, but it isn't necessarily so. How do shadows show the sun's apparent motion? How can you use Earth's rotation to explain shadow movement?

It's Apparent

8

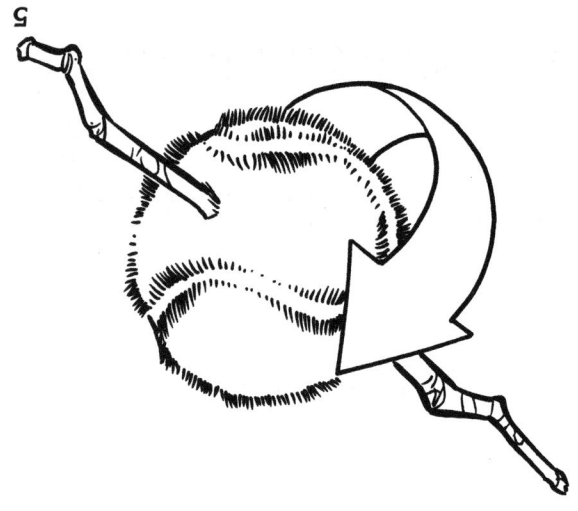

However, the Earth's axis is not visible like the stick.

- The Earth spins on its axis, much like the tennis ball spins on the stick.
- Earth is a sphere, so is the tennis ball.

The tennis ball and stick make a pretty good model in which to investigate Earth's rotation.

This makes sense! It would be easy for us to agree with them if we didn't have the work of scientists who discovered that it is the Earth that is actually moving. It spins. This spinning action is called rotation. To help you understand, imagine poking a sharp stick through a tennis ball and then spinning the ball.

- You can even position the tennis ball to be tilted like the Earth's axis. It is tilted a little more than 23 degrees. That would be like using the stick to represent an hour hand of a clock and moving the hour hand from 12 o'clock to not quite 1 o'clock.

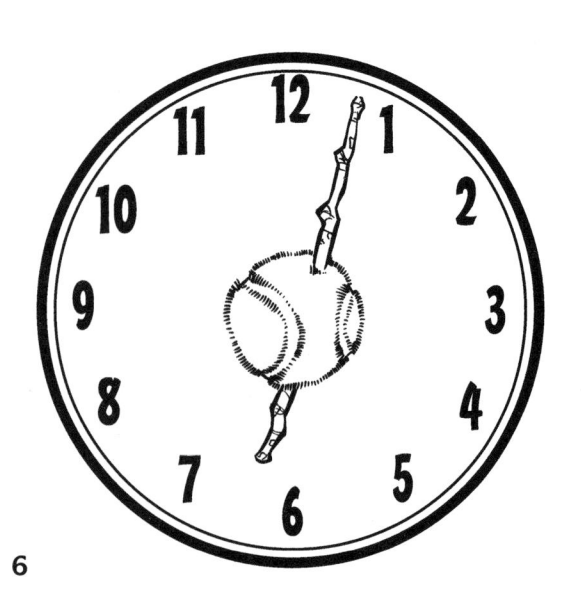

It seems like the sun moves, doesn't it? That's what early astronomers thought. They thought that the sun went around the Earth. They saw the sun in the eastern sky in the morning. At noon it was overhead, and toward evening it was in the west.

0033

OUT OF THIS WORLD

0010010101101001001010101

SPIN CYCLE

001001010110100100101

0-7041-776

Topic
Earth's rotation

Key Question
In what direction does the Earth rotate?

Learning Goals
Students will:
- make and use a model to discover the direction of Earth's rotation, and
- apply their learning to the need for time zones.

Guiding Documents
Project 2061 Benchmarks
- *Like all planets and stars, the earth is approximately spherical in shape. The rotation of the earth on its axis every 24 hours produces the night-and-day cycle. To people on earth, this turning of the planet makes it seem as though the sun, moon, planets, and stars are orbiting the earth once a day.*
- *Use numerical data in describing and comparing objects and events.*

NRC Standard
- *Most objects in the solar system are in regular and predictable motion. These motions explain such phenomena as the day, the year, phases of the moon, and eclipses.*

NCTM Standards 2000*
- *Recognize the attributes of length, volume, weight, area, and time*
- *Use representations to model and interpret physical, social, and mathematical phenomena*

Math
Measurement
    time zones

Science
Earth science
    rotation

Integrated Processes
Observing
Comparing and contrasting
Applying

Materials
Tennis balls
Adhesive labels (see *Management 1*)
Yellow construction paper or flashlights
2 pushpins
Globe

Background Information
A *day* is the time it takes for Earth to make one complete rotation about its axis, about 24 hours. The direction of the Earth's spin is counterclockwise when viewed looking down on the North Pole. Because the sun is stationary while the Earth rotates, most locations have a period of daylight and a period of darkness during these 24 hours.

To the observer on Earth, it appears that the sun is moving across the sky from sunrise (east) to noon to sunset (west), causing day and night. Students need multiple experiences to be convinced that it is the Earth that is moving rather than the sun.

In this experience, students will spin a tennis ball globe to determine the direction of spin that causes the sun to appear in the eastern sky and set in the western sky. This activity will then be related to time zones within the contiguous United States.

Management
1. Copy the maps onto adhesive labels. Use labels that are 2" x 4 1/4" such as Avery #10.
2. Yellow construction paper can be used to design a sun, or a student within each group can hold a flashlight to shine on the model of the Earth. The flashlight serves as a better representation of the sun.
3. Have students work together in groups of three or four.
4. The globe is used as a reference for placement of the map. It can also be used to explore day and night concepts for the world as well as time zone information.

Procedure
1. Ask the *Key Question* and state the *Learning Goals*.
2. Tell the students that they are going to make a model of the Earth to discover the answer to the question. Distribute the construction sheet, the tennis balls, string, adhesive labels, and the pushpins.

21

3. Inform the students that the pushpins will represent the Earth's axis. Have them put one pin in the tennis ball and the other pin at the opposite end. Let them put their fingertips on the pushpins and twirl the tennis balls.

4. Ask students to cut out the map of the contiguous United States, peel off the backing, and position it on the tennis ball. Have them use the globe as a reference for the placement of the map.

5. Either place the large sun cutout on a wall or distribute flashlights to each group. Ask the students the direction in which the sun appears. Then remind them that the sun appears in the east because of the Earth's rotation.

6. Tell them that they are to try to figure out which way the Earth rotates in order for the sun to appear in the east first and to set in the west. Point out that the letters representing the directions have been placed on the map.

7. After a period of investigation, hold a time of discussion that results in students concluding that the Earth rotates in a counterclockwise direction when viewed looking down on the North Pole.

8. Next draw the students' attention to the somewhat vertical lines that have been drawn on the map. Tell the students that these represent time zones. Each adjacent time zone represents one hour difference in time. Starting on the east coast and moving west, point out that the time decreases one hour with each area. Give students the example that it is 8 o'clock in the morning along the east coast. At the same time in the next time zone to the west, it is 7 o'clock, 6 o'clock in the next, and 5 o'clock along the west coast. Ask why they think this is the case.

9. Make sure that students understand that time zones exist all over the world and not just in the United States.

10. Distribute the activity page and allow time for students to complete it.

Connecting Learning

1. In what direction does the Earth rotate? How do you know?
2. What would happen if it rotated the other direction?
3. How long does it take for the Earth to make one rotation?
4. What would happen if the time of rotation were different?
5. Why do you think we have time zones?
6. Is the time of day in New York City earlier or later than in San Francisco? Explain how you know.
7. Often on television, it will say that a program is airing at 8:00 eastern, 7:00 central. How does this apply to this activity?
8. Sporting events are often shown live. If a game is being played in San Diego at 2:00 P.M., at what time would it be showing in Dallas? ...in Boston?
9. What are you wondering now?

Internet Connections

http://www.travel.com.hk/region/timezone.htm

This is a very colorful world time zone map. Students will see some discrepancies in the pattern, but should be able to generalize that there are 24 time zones around the world.

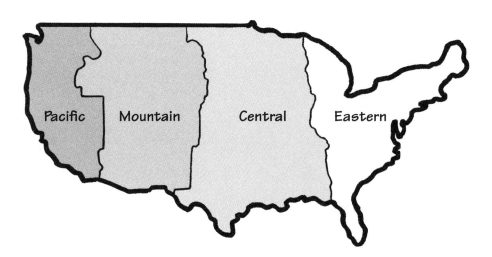

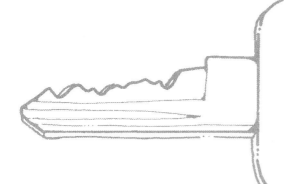

Key Question

In what direction does the Earth rotate?

Learning Goals

Students will:

- make and use a model to discover the direction of Earth's rotation, and

- apply their learning to the need for time zones.

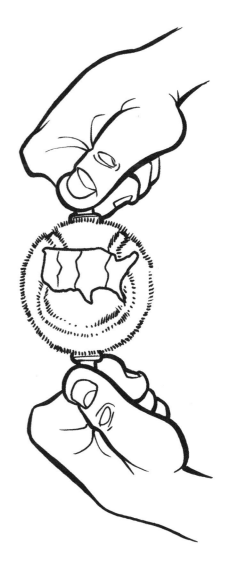

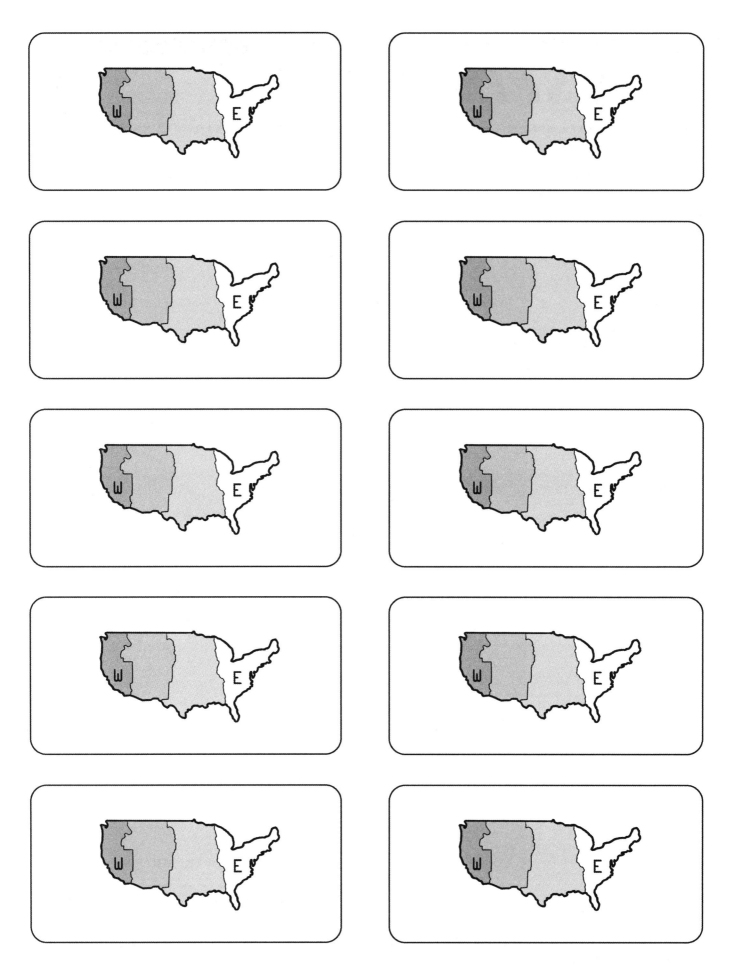

SPIN CYCLE
TENNIS BALL EARTH CONSTRUCTION   0010010101101001001010   0-7041-776

Tennis ball
2 pushpins
Map label

1. Stick one pushpin in the tennis ball to represent the north pole.

2. Stick another pushpin in the opposite end of the tennis ball to represent the south pole.

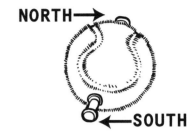

NORTH→

←SOUTH

3. Cut out the map of the contiguous United States. Peel off the backing and stick the label on the tennis ball. Look at a globe for placement of the label.

4. Grasp the pushpins and twirl the Earth. Find out which direction the Earth rotates.

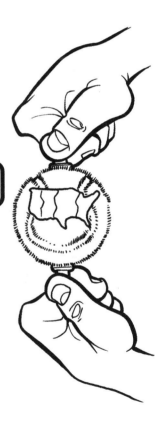

Use these words to fill in the blanks:

| **East** | **West** | **Earth's axis** | **Sphere** | **Rotates** |

1. The two pushpins represent the _____.

2. The Earth _____ once every 24 hours.

3. The E on the map stands for _____. The sun

   appears to rise in the _____.

4. The W on the map stands for _____. The sun

   appears to set in the _____.

Circle the right answer(s):

The Earth rotates

    a. on its axis.   b. upside down.   c. every 24 hours.

    d. counterclockwise.   e. clockwise.   f. only once a year.

SPIN CYCLE

1. In which direction does the Earth rotate?

2. How do you know?

3. There are four time zones in the contiguous United States. Because the sun first appears in the east, each time zone is one hour behind the time zone to its east. Here is a map of the four time zones. In the illustration, a clock is set in one time zone. Set the clocks in the other time zones accordingly.

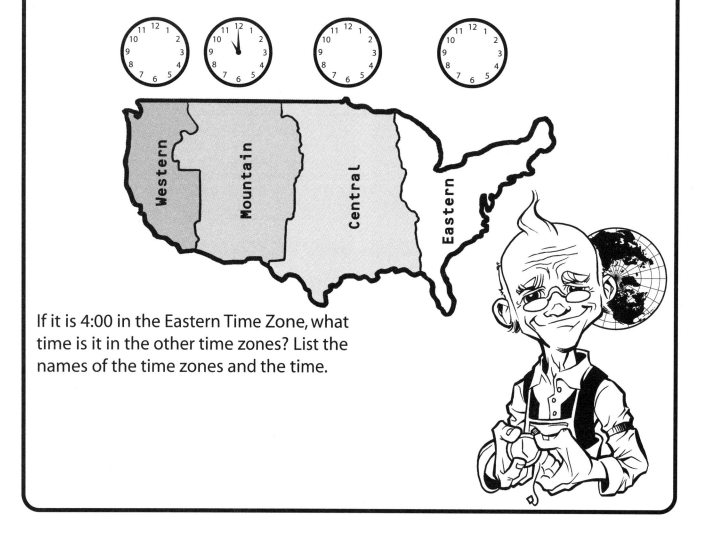

If it is 4:00 in the Eastern Time Zone, what time is it in the other time zones? List the names of the time zones and the time.

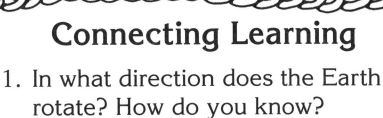

SPIN CYCLE

Connecting Learning

1. In what direction does the Earth rotate? How do you know?

2. What would happen if it rotated the other direction?

3. How long does it take for the Earth to make one rotation?

4. What would happen if the time of rotation were different?

5. Why do you think we have time zones?

6. Is the time of day in New York City earlier or later than in San Francisco? Explain how you know.

7. Often on television, it will say that a program is airing at 8:00 eastern, 7:00 central. How does this apply to this activity?

8. Sporting events are often shown live. If a game is being played in San Diego at 2:00 P.M., at what time would it be showing in Dallas? ...in Boston?

9. What are you wondering now?

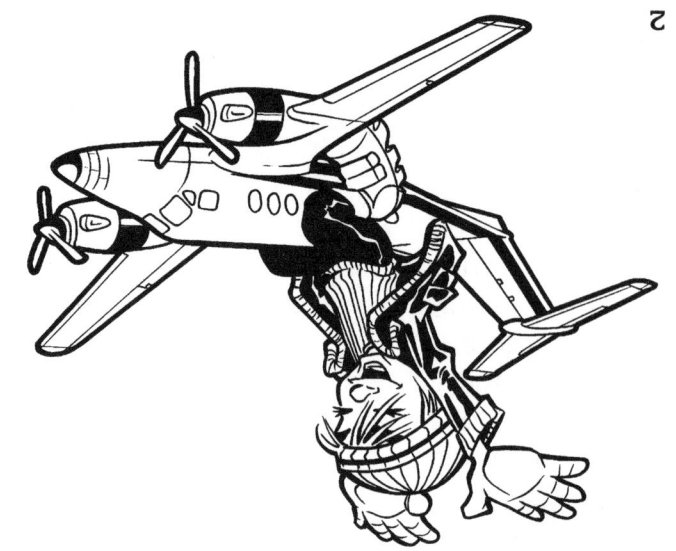

A major airline has a flight leaving San Francisco, California at 6:00 A.M and arriving in Salt Lake City, Utah at 8:42 A.M. That same airline has a flight leaving Salt Lake City, Utah at 3:00 P.M and arriving in San Francisco, California at 3:50 P.M. Both flights are non-stop. How do you explain the differences in flight times?

The time zones in the United States are defined in the U.S. Code within a chapter called *Standard Time*. The Department of Transportation has authority over the time zones in the United States. Over time they have adjusted the time zones and renamed them.

The United States and its territories use eight time zones. Here is a chart that relates these time zones to Universal Time.

Time Zone	Compared with Universal Time
Atlantic Standard Time	UT–04:00
Eastern Standard Time	UT–05:00
Central Standard Time	UT–06:00
Mountain Standard Time	UT–07:00
Pacific Standard Time	UT–08:00
Alaska Standard Time	UT–09:00
Hawaii-Aleutian Standard Time	UT–10:00
Samoa Standard Time	UT–11:00

12

Time zones explain the differences! San Francisco is in the Pacific Time Zone. Salt Lake City is in the Mountain Time Zone. To fly from San Francisco to Salt Lake City, a passenger must set his or her watch ahead one hour. Instead of leaving at 6:00, the passenger is actually leaving at 7:00 Salt Lake City time. All the calculations must be adjusted to fit the time zones in which the flights occur.

In order to have clear communication with people in other areas, you may need to know time zone information. Some people—astronomers, navigators, etc.—report their time using Universal Time (UT). This is the time taken from the zone that passes through Greenwich, England. It serves as an official reference point and is maintained by a large number of very accurate "atomic clocks."

All one has to do to report their time in UT is to say how many time zones it is from Greenwich. For time zones east of Greenwich, they are reported with a + sign. For time zones to the west of Greenwich, a – sign is used. The Eastern Time Zone would report theirs as UT–05:00 because it is 5 time zones to the west of Greenwich.

A further check of the flight information says that the first flight takes one hour and 42 minutes. The returning flight takes one hour and 50 minutes. Do your calculations give you those results? What would explain these differences?

IS THIS GREENWICH? YEAH, I'M CLOCKING IN AT -5:00. DO YOU CONFIRM?

IT'S ALL VERY CONFUSING.

I DON'T EVEN KNOW. NO ONE KNOWS, REALLY.

WHERE I GO YOU CAN'T FOLLOW... WHEN I GET THERE... NO ONE KNOWS.

With goods and people going many places, it became necessary to communicate information like when they would arrive. And that's where the problem came in. Up until this time, people could set their clocks however they wanted to. There were hundreds of different times adopted in different places in the world. Things needed to be simplified.

Sea navigation became faster and ships were sailing to many new ports around the world. The railroad system expanded to more and more areas.

The Earth's surface was divided into 24 adjacent wedges above and below the equator. Each wedge would be one-hour difference in time. Going around the entire world, through the 24 wedges would equal 24 hours, or one day. Some locations still adopt a time different from their time zone, but most use the time zone given to them.

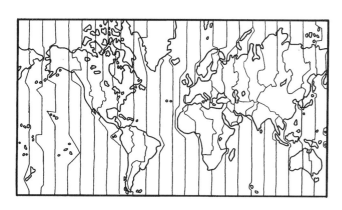

So why do we have time zones? Time zones didn't even exist until the 1880s. What happened in this period of time that created the need for time zones? Transportation was developing at a very rapid rate.

Y'KNOW WHAT WOULD MAKE TIME GO BY FASTER?

YEP. TIME ZONES.

LUNAR LOOKING

Topic
Phases of the moon

Key Question
What changes does the moon go through each month?

Learning Goals
Students will:
- identify the stages of the moon's cycle, and
- sequence the stages in the correct order.

Guiding Documents
Project 2061 Benchmarks
- *The sun can be seen only in the daytime, but the moon can be seen sometimes at night and sometimes during the day. The sun, moon, and stars all appear to move slowly across the sky.*
- *The moon looks a little different every day, but looks the same again about every four weeks.*

NRC Standard
- *The sun, moon, stars, clouds, birds, and airplanes all have properties, locations, and movements that can be observed and described.*

Science
Earth science
   astronomy
      moon phases

Integrated Processes
Observing
Classifying
Sequencing

Materials
For the class:
   moon model transparency
   The Moon Book (see *Management 3*)
   glue

For each student:
   set of moon phase cards
   set of turn around pages
   scissors

Background Information
The moon is the brightest light visible in our night sky, yet it produces no light of its own—it only reflects light given off by the sun. The sun is a star, and like other stars is composed of burning gases that emit light. This light reflects off the moon and other planets in our solar system, allowing us to see them at night. The phases of the moon are a result of the moon's position relative to the sun as it orbits the Earth.

The moon is said to be a *new moon* when its orbit places it between the sun and the Earth. The next several nights, the slice of the moon that is visible is called a crescent moon. It is called a waxing crescent until the moon has completed one-quarter of its 28-day orbit. Even though at this point one-half of the moon is visible to us, the phase is called the first quarter moon. This is because the moon has traveled one-quarter of the way around the Earth. As the moon continues its orbit between the first quarter and full moon stages, it is called a waxing gibbous moon. The moon is considered *full* when it has traveled half way around the Earth. At this point, the Earth is between the moon and the sun, so we see a full circle illuminated in the night sky. The moon is visible at this stage because of the relative positions of the sun, moon, and Earth. Over the next several days, the moon is in the gibbous stage and is said to be waning. When the moon has traveled three-quarters of the way around the Earth, it is called the last quarter moon. The last few days of the moon's orbit are called the waning crescent moon. When the moon has completed its orbit, it becomes a new moon once more and the cycle continues.

Management
1. Use the page provided to make a transparency strip showing the moon phases for use during class discussion.
2. Prepare a set of eight moon phase cards for each student. Copy on card stock and laminate for extended use.
3. You will need a copy of *The Moon Book* by Gail Gibbons. (Scholastic, Inc. New York. 1997.)

Procedure

Part One

1. Read and discuss the pages from *The Moon Book* that address the phases of the moon.
2. Point out to students that the moon's shape each night follows a pattern. Use the transparency model to look at the pattern in the phases of the moon.
3. Distribute one set of moon phase cards to each student.
4. Direct the students to sequence the eight cards as you review the transparency strip. Discuss with the students the proper names for each of the phases.
5. The focus of this learning experience is for the students to learn that the phases of the moon follow a specific sequence.

Part Two

1. Divide the students into groups of four.
2. Have each group combine their sets of moon phase cards to create a four-set deck.
3. Direct one of the students to shuffle the cards and deal eight cards to each student.
4. Explain that the object of the game is to collect a complete eight-card sequence of the phases of the moon.
5. Tell the students that play begins with the student to the left of the dealer. That student must randomly draw a card from the hand of any other player, and then replace it with any card from his/her hand.
6. Play continues in a clockwise manner until someone completes a moon sequence.

Part Three

1. Give each student a copy of the turn-around pages and have them cut along the bottom of the first page as indicated.
2. Assist them in lining up the two pages and gluing them together as described on the second page.

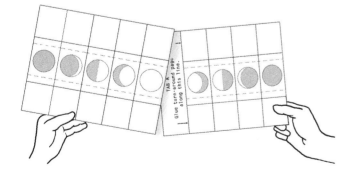

3. Instruct students to orient their papers horizontally on their desks and fold the tops of the pages under along the dashed lines.

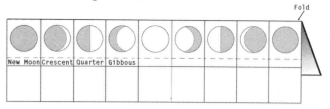

4. Assist the students in labeling each of the phases of the moon directly beneath the illustrations. Ask the students to use the spaces provided below the names to describe those moon phases. Encourage them to discuss what the pictures show and what causes us to see each phase of the moon in the sky.
5. When the students have finished labeling the stages, question them about what would come next in the sequence of the moon phases. (It may be necessary to revisit the transparency strip used in *Part One* of this activity at this time.) Guide students to discover that the 10th picture would be the same as the second, the 11th would be the same as the third, and so on.
6. Following this discussion, have students unfold their papers, turn them upside down, and fold under the top portion as they did in *Procedure 3*. Have them repeat the process described in *Procedure 4*.
7. When students have labeled both halves of their turn-around pages, discuss what they have discovered about the predictable pattern of the moon in the night sky.

Connecting Learning

1. How does the moon look when it is full?
2. Since it takes about 28 days to complete a moon cycle, how many cards would you need to add to our set of eight cards to illustrate the entire moon cycle? [20]
3. What would the cards look like?
4. Which moon phase would apply to us today?
5. Do you always see the moon at night? Explain.
6. Have you ever seen the moon during the day?
7. What other things in the sky occur in a sequence or cycle? [sun rising and setting]
8. Why is it important to look for sequences and cycles? [Many things in nature occur in patterns.]
9. How can you tell the difference between a first quarter moon and a last quarter moon? [by which side is illuminated]
10. What are you wondering now?

0028

OUT OF THIS WORLD

00100101011010010010101010

LUNAR LOOKING

00100101011010010010101010

03-182-005

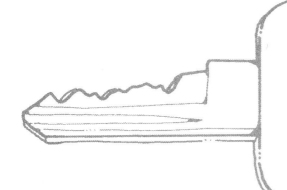

Key Question

What changes does the moon go through each month?

Learning Goals

Students will:

- identify the stages of the moon's cycle, and

- sequence the stages in the correct order.

| 1. New Moon | 2. Waxing Crescent | 3. Waxing Crescent | 4. Waxing Crescent |

| 5. First Quarter | 6. Waxing Gibbous | 7. Waxing Gibbous | 8. Full Moon |

| 9. Waning Gibbous | 10. Waning Gibbous | 11. Waning Gibbous | 12. Last Quarter |

| 13. Waning Crescent | 14. Waning Crescent | 15. Waning Crescent | 16. New Moon |

Make a transparency of this sheet. Cut each strip apart, then tape the four strips together to make one long strip. Use an index card to make a holder. Pull the transparency through on the overhead to show moon phases.

Lunar Looking Moon Phase Cards

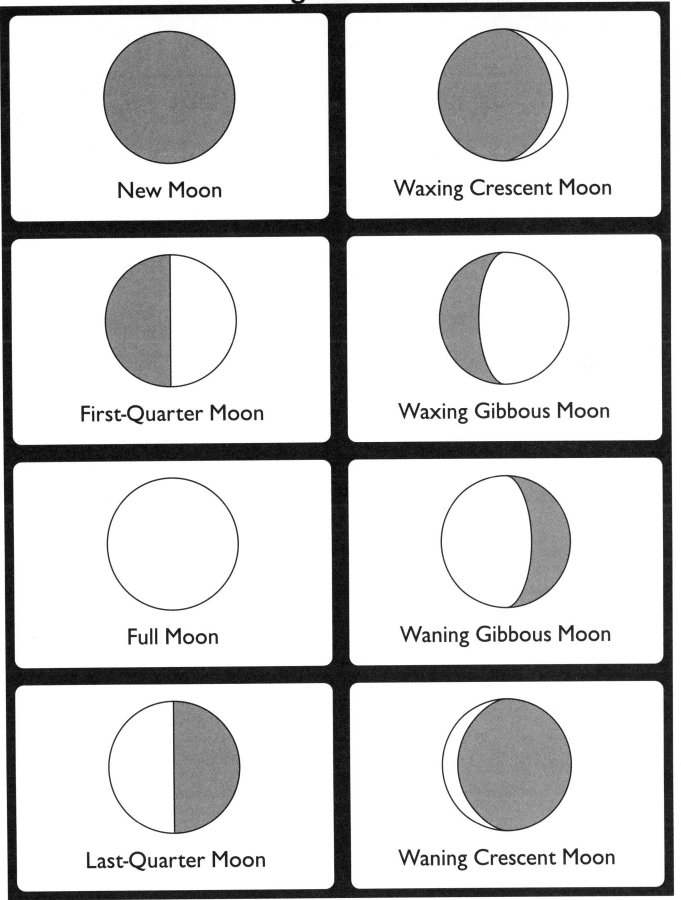

New Moon

Waxing Crescent Moon

First-Quarter Moon

Waxing Gibbous Moon

Full Moon

Waning Gibbous Moon

Last-Quarter Moon

Waning Crescent Moon

Lunar Looking Turn-Around Page One

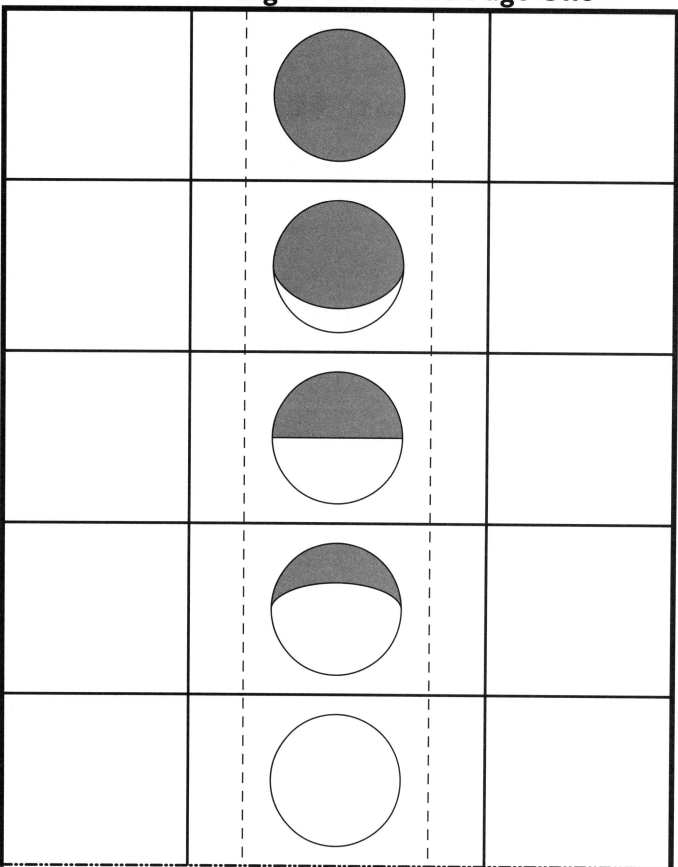

Cut along all outside lines. Glue this page to tab A on the following page. Align the matching lines.

Lunar Looking Turn-Around Page Two

Tab A

Cut along the outside lines. Align the matching line from page one. Glue the two pages together.

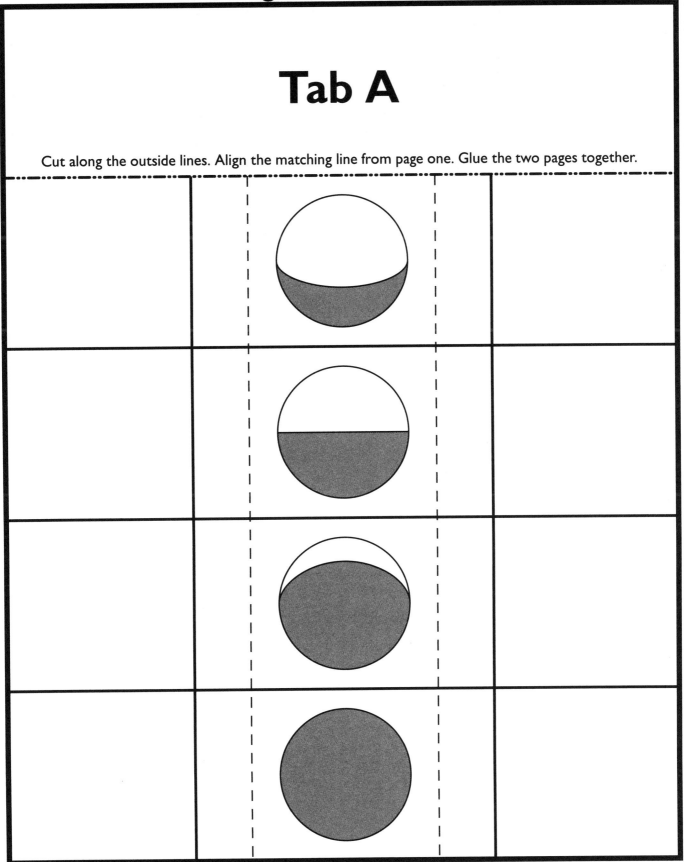

LUNAR LOOKING

Connecting Learning

1. How does the moon look when it is full?

2. Since it takes about 28 days to complete a moon cycle, how many cards would you need to add to our set of eight cards to illustrate the entire moon cycle?

3. What would the cards look like?

4. Which moon phase would apply to us today?

5. Do you always see the moon at night? Explain.

LUNAR LOOKING

Connecting Learning

6. Have you ever seen the moon during the day?

7. What other things in the sky occur in a sequence or cycle?

8. Why is it important to look for sequences and cycles?

9. How can you tell the difference between a first quarter moon and a last quarter moon?

10. What are you wondering now?

Topics
Moon, Earth, Sun relationships
Moon phases

Key Question
What causes the changing moon phases?

Learning Goals
Students will:
- construct a moon-Earth model where they take the position of an Earth-bound observer, and
- use the model to determine how and in what order the phases of the moon occur.

Guiding Documents
Project 2061 Benchmarks
- *The moon's orbit around the earth once in about 28 days changes what part of the moon is lighted by the sun and how much of that part can be seen from the earth—the phases of the moon.*
- *Different models can be used to represent the same thing. What kind of a model to use and how complex it should be depends on its purpose. The usefulness of a model may be limited if it is too simple or if it is needlessly complicated. Choosing a useful model is one of the instances in which intuition and creativity come into play in science, mathematics, and engineering.*
- *Things that change in cycles, such as the seasons or body temperature, can be described by their cycle length or frequency, what the highest and lowest values are, and when they occur. Different cycles range from many thousands of years down to less than a billionth of a second.*

NRC Standard
- *The sun, moon, stars, clouds, birds, and airplanes all have properties, locations, and movement that can be observed and described.*

NCTM Standards 2000*
- *Describe location and movement using common language and geometric vocabulary*

- *Recognize geometric ideas and relationships and apply them to other disciplines and to problems that arise in the classroom or in everyday life*

Math
Spatial visualization
Geometric modeling

Science
Earth science
   astronomy
      moon phases

Integrated Processes
Observing
Comparing and contrasting
Collecting and recording data
Interpreting data
Applying

Materials
For moon collars:
   foam core board (see *Management 5*)
   craft knife
   black permanent marker
   Moon Collar Guides (see *Management 6*)
   meter stick
   8 jumbo paper clips
   two sharpened pencils

For each group:
   moon collar
   table tennis ball (see *Management 2*)
   Alignment Square
   modeling clay
   tape
   scissors
   Moon Phase Pictures, one set
   Using the Moon Collar page

For each student:
   Moon Phase Pictures, one set
   student pages

Background Information

The moon orbits the Earth. It takes about a month for the moon to complete its orbit. Its orbit is nearly circular with a radius of about 380,000 km (228,000 mi). The moon revolves in a counterclockwise direction as viewed from the perspective of the Earth's north pole.

The moon produces no light of its own. It only reflects light. One half of the moon is always lit by sunlight. The dark half is shaded from the sun by the moon itself. The portion of the moon that appears to be lit by the sun changes for the Earth-bound observer. This change is the result of the changing relative positions of the sun, moon, and Earth.

When the moon is between the sun and the Earth, the lit side is facing the sun and the shadowed side is facing the Earth. Since we cannot see any lit part of the moon, it is not visible to us even while it is in the sky. This phase is called a *new moon*. When the Earth is between the moon and the sun, the entire lit side of the moon is facing the Earth. This phase is a *full moon*. The other phases of the moon fall between these extremes with apparently different amounts of the moon lit.

The naming of the phases is not the emphasis of this activity, but it allows students to learn scientific vocabulary. The retention of the vocabulary is better if there is an understanding of the terms. *Waxing* refers to increasing in size. One way to help students remember the definition is to refer to the increase in size of candles as they are dipped in wax. *Waning* means the decline of something, in this case the apparently diminishing light of the moon. *Gibbous* finds its root in a Latin word for hump. The *quarters* are a division of the monthly cycle of the moon into four equal periods. A full moon might be referred to as the second quarter.

Students generally incorrectly conceive that the phases of the moon are caused by the shadow of the Earth covering part of the moon. A correct understanding of what causes the phases of the moon can be understood as the students observe the regular changes of the moon's appearance, and the moon's relative position to the sun, as well as developing models of the geometry of the situation, and experiences with the situation's scale.

This activity addresses **only** the modeling of the situation. Students should recognize the moon is a sphere that can only be half lit by any single light source. The model will allow them to recognize that the amount of the lit half of a sphere that is observed depends on the observer's position relative to the sphere and the light source. By connecting the model with dated observations of the moon, students can conclude that the regular change in the moon's appearance is caused by the constantly changing relative positions of the moon, sun, and Earth-bound observer as the moon consistently revolves around the Earth.

Students using this model often have a hard time understanding how the full moon could exist on the far side of the Earth relative to the sun. In this position they argue the sun's light should be blocked and the moon should be in Earth's shadow. (This is what happens during a lunar eclipse). This is a problem of scale that this model does not address. At the scale of Earth and the moon, the shadow of the Earth is very narrow. On the moon's revolution, the arc of the shadow is less than a degree. The consequence is that the moon would only spend 1/360 of its revolution (less than two hours) in this shadow. To complicate the problem, the moon's revolution is not in the same plane as the Earth's revolution. This means that there are only two times a year when the Earth is between the sun and the moon to cause the shadow **and** the planes of revolution intersect. This is why lunar eclipses generally happen only twice a year.

Management

1. Students should have observed the moon for four to six weeks before doing this activity so that they are aware that the moon changes in regular sequence. They will need to make a record of the dates of observations and the appearance of the moon for these dates. If student data are not feasible, find a calendar or newspaper record of the phases of the moon for the period around the time of the activity.

2. Before doing this activity, color half of each of the table tennis balls with a black permanent marker. This will represent the moon's lit and shadowed sides.

3. Students need to work together in groups of five.

4. Each group will need a large piece (70 cm x 70 cm minimum) of foam core board for its moon collar. Foam core board can be found at arts and crafts or teacher supply stores. Large sheets of sturdy cardboard can also be used.

5. It is recommended that you prepare the moon collars for groups ahead of time, but if desired, they can assist in the construction. Allow additional time for this process.

6. Make a transparency copy of the *Moon Collar Guides* page.

7. Each group will need one set of the *Moon Phase Pictures* (eight pictures), and each student will also need his or her own set. Each group needs only one *Alignment Square*. The page of *Alignment Squares* should be copied onto card stock, if possible.

Procedure

1. Discuss the *Key Question* with the students to assess the students' understanding about the phases of the moon.

2. Divide students into groups of five and distribute the necessary materials to each group.

3. Read through the instructions for using a moon collar together as a class to be sure that everyone understands the procedure and the responsibilities of the four roles.

4. Allow time for students to complete the process of viewing the moon model at the different phases. Have groups repeat the process with different students taking on the role of *Moon Viewer* so that everyone can have a chance to see the changes.

5. Distribute the additional set of *Moon Phase Pictures* to each student along with the two student pages.

6. Have students use the *Moon Phase Pictures* to make their own models of the moon's phases and label the corresponding dates.

7. Once students have answered the questions on the final page, discuss what they observed and learned from this activity.

Connecting Learning

1. How much of the moon is always lit by the sun? [one-half]

2. Determine which way the moon is revolving around the Earth and explain how you know. [counterclockwise from north pole, sequence of dates]

3. Where must the moon be relative to the Earth and the sun to appear to be fully lit? [on the opposite side of the Earth from the sun; sun →Earth →moon] Does this seem logical? Explain.

4. Why does the moon appear to be fully lit when the sun is shining on only half the moon? [can see only one-half at a time, lit half towards Earth]

5. Where must the moon be relative to the Earth and the sun to appear to be half lit? [perpendicular to line between sun and Earth]

6. How much of the whole moon are you seeing when half of the moon appears to be lit? [one-quarter]

7. Why is a moon that appears to be half lit called either a first quarter or third quarter? [first and third quarter of sequence of phases]

8. Where must the moon be relative to the Earth and the sun to appear to be a crescent? [at an acute angle to the line between sun and Earth]

9. Why does so little of the moon appear to be lit in this position? [most of lit half is facing away from Earth]

10. Where must the moon be relative to the Earth and the sun to appear to be a gibbous (hump-backed)? [at an obtuse angle from line between sun and Earth]

11. Why are you not able to see the whole lit half of the moon in this position? [part of lit half is still hidden behind the moon]

12. What are you wondering now?

Extension

Have students observe the moon as a waxing crescent and see if they can see the dark half of the moon lit by Earthshine. Ask them to use the model to explain why the dark side of the moon is visible in this phase.

* Reprinted with permission from *Principles and Standards for School Mathematics,* 2000 by the National Council of Teachers of Mathematics. All rights reserved.

FACING UP TO THE MOON

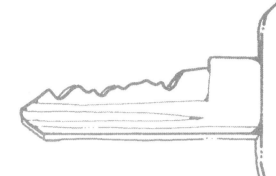

Key Question

What causes the changing moon phases?

Learning Goals

Students will:

- construct a moon-Earth model where they take the position of an Earth-bound observer, and

- use the model to determine how and in what order the phases of the moon occur.

FACING UP TO THE MOON

MOON COLLAR GUIDES · 00100101011010010010101 · 091-120-01

Copy this page onto transparency film. These are your moon collar guides. Use the guides to mark and then outline the eight locations of the *Alignment Square* on the moon collar.

1. Find and mark the center of the piece of foam board.

2. Link three jumbo paper clips together and use them as a compass to draw a circle in the center of the foam board.

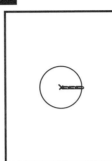

3. Link four or five additional paper clips to your chain and use them as a compass to draw a larger circle around the one you just drew. (Be sure the circle is at least 3 cm from all edges of the foam board.)

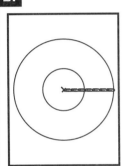

4. Place the transparency of the *Moon Collar Guides* page on the foam board so that the center dot of the circle is at the center of the circles you drew. (The edges of page should be parallel to the edges of the foam board.)

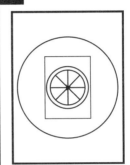

5. Lay a meter stick along each of the lines in turn and mark the places where the stick intersects with the larger, outer circle. This should give you eight dots all equally spaced around the circumference.

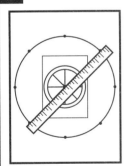

6. Using a craft knife, cut out the smaller circle in the center of the foam board.

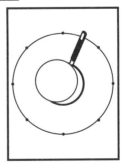

7. Cut out the square from the *Moon Collar Guides* page. Align its center with each of the dots you made on the circumference of the large circle. Trace around the square at each position using the permanent marker. Be sure that the sides of the square are parallel to the sides of the foam board.

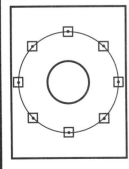

8. Draw several parallel lines along the length of the foam board using the permanent marker. Put arrows at one end of the lines to indicate the direction of the sunlight.

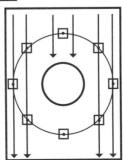

FACING UP TO THE MOON

USING THE MOON COLLAR   001001010110100100101⸱

091-120-01

You Need:

Moon collar   Table tennis ball   *Alignment Square*   Clay
Tape   Scissors   *Moon Phase Pictures*

Roles:

Rotate roles until everyone has a chance to be the *Moon Viewer*.

Moon Viewer: Has head in moon collar to view moon in various phases
Moon Mover: Moves the moon (table tennis ball) around the moon collar
Phase Fixer: Holds up the moon phase cards and puts them in the proper locations
Collar Holders (2): Hold the moon collar in place

1. Form a ring from a small snake of clay. Use the ring to attach the table tennis ball to the *Alignment Square*. Match the dark side of the ball with the square's dark side.

2. Cut apart one set of eight *Moon Phase* pictures and give them to the *Phase Fixer*.

3. *Moon Viewer:* Put your head up through the hole in the foam board while the *Collar Holders* hold it in place.

4. *Moon Mover:* Place the table tennis ball with the *Alignment Square* in one of the outlined squares on the large circle. The direction of the sunlight arrows on the square must match the direction of the arrows on the moon collar.

5. *Moon Viewer:* Turn and look at the table tennis ball to see what "phase" it is in.

6. *Phase Fixer:* Hold up each of the *Moon Phase Pictures* to show the *Moon Viewer*. When the *Moon Viewer* tells you which phase he/she sees, place that picture onto the moon collar next to the *Alignment Square*.

7. Repeat this process for each of the positions marked on the circle. The *Moon Viewer* should turn to face the table tennis ball each time it is moved.

8. When all the *Moon Phase Pictures* have been placed, record the calendar dates that correspond to when the moon looked like each picture.

FACING UP TO THE MOON

Copy this page onto card stock. Each group needs one square.

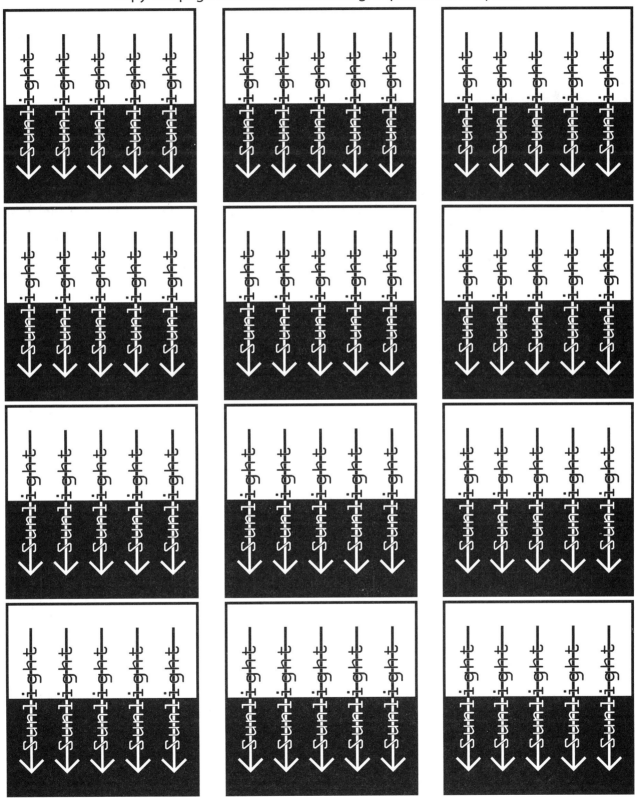

0025

00100101011010010101010

OUT OF THIS WORLD

FACING UP TO THE MOON

MOON PHASE PICTURES   00100101011010010101010

091-120-01

Each group needs one set of pictures. Each student needs an additional set.

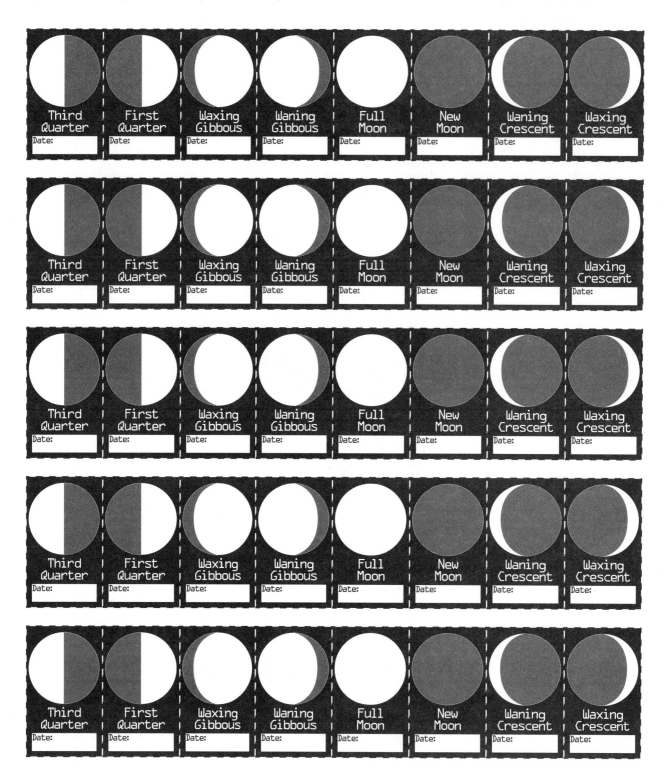

FACING UP TO THE MOON

Cut out the moon phase pictures. Place them in the appropriate positions with the date end nearest the Earth. Tape or glue the pictures in place. Record the dates for each "phase."

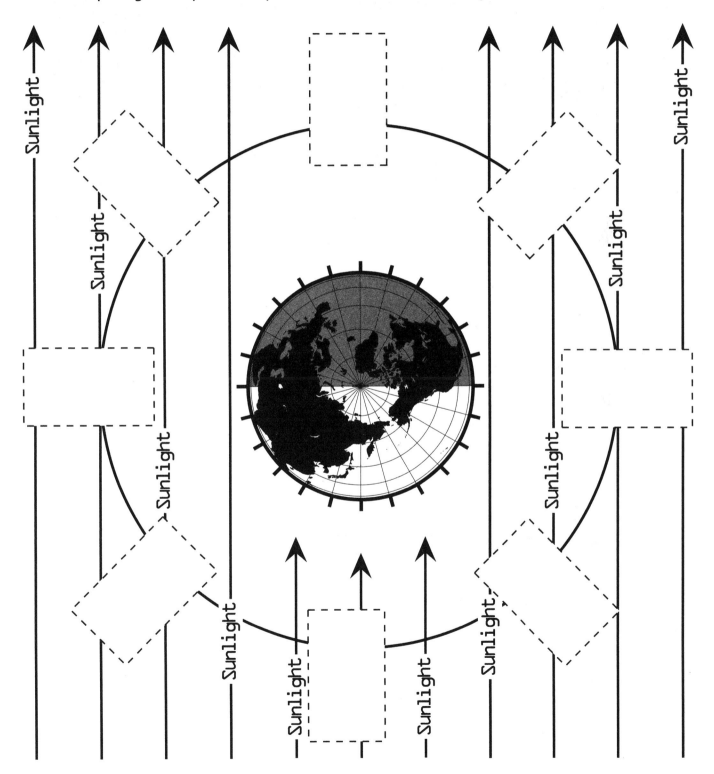

49

FACING UP TO THE MOON

1. How long does it take for the moon to orbit the Earth?

2. How much of the moon is lit when we say we see a full moon?

3. What are the positions of the Earth, sun, and moon when there is a new moon?

4. Why is a moon that is half lit called either a first quarter or a third quarter?

5. What fractions apply to the phases of the moon?

Connecting Learning

1. How much of the moon is always lit by the sun?

2. Determine which way the moon is revolving around the Earth and explain how you know.

3. Where must the moon be relative to the Earth and the sun to appear to be fully lit? Does this seem logical? Explain.

4. Why does the moon appear to be fully lit when the sun is shining on only half the moon?

5. Where must the moon be relative to the Earth and the sun to appear to be half lit?

6. How much of the whole moon are you seeing when half of the moon appears to be lit?

0025

00100101011010010010101010

OUT OF THIS WORLD

FACING UP TO THE MOON

00100101011010010010101010

091-120-01

Connecting Learning

7. Why is a moon that appears to be half lit called either a first quarter or third quarter?

8. Where must the moon be relative to the Earth and the sun to appear to be a crescent?

9. Why does so little of the moon appear to be lit in this position?

10. Where must the moon be relative to the Earth and the sun to appear to be a gibbous?

11. Why are you not able to see the whole lit half of the moon in this position?

12. What are you wondering now?

THE MOON SHINES BRIGHT

0010010101101001001010

0-3081-984

Topic
Lunar cycles

Key Question
What can you learn about the movement of the moon by observing it at different times during the day?

Learning Goals
Students will:
- observe and record the location of the moon in the sky at the same time on two consecutive days, and
- use the information to describe the cycle of the moon and how its location changes from day to day.

Guiding Documents
Project 2061 Benchmarks
- *Like all planets and stars, the earth is approximately spherical in shape. The rotation of the earth on its axis every 24 hours produces the night-and-day cycle. To people on earth, this turning of the planet makes it seem as though the sun, moon, planets, and stars are orbiting the earth once a day.*
- *Graphical display of numbers may make it possible to spot patterns that are not otherwise obvious, such as comparative size and trends.*
- *Things change in steady, repetitive, or irregular ways—or sometimes in more than one way at the same time. Often the best way to tell which kinds of change are happening is to make a table or graph of measurements.*
- *The graphic display of numbers may help to show patterns such as trends, varying rates of change, gaps, or clusters. Such patterns sometimes can be used to make predictions about the phenomena being graphed.*
- *Things that change in cycles, such as the seasons or body temperature, can be described by their cycle length or frequency, what the highest and lowest values are, and when they occur. Different cycles range from many thousands of years down to less than a billionth of a second.*
- *Organize information in simple tables and graphs and identify relationships they reveal.*

NRC Standards
- *Most objects in the solar system are in regular and predictable motion. Those motions explain such phenomena as the day, the year, phases of the moon, and eclipses.*

- *Use appropriate tools and techniques to gather, analyze, and interpret data.*
- *Develop descriptions, explanations, predictions, and models using evidence.*
- *Use mathematics in all aspects of scientific inquiry.*

NCTM Standards 2000*
- *Collect data using observations, surveys, and experiments*
- *Represent data using tables and graphs such as line plots, bar graphs, and line graphs*
- *Propose and justify conclusions and predictions that are based on data and design studies to further investigate the conclusions or predictions*

Math
Data analysis
   line graph
Measurement
   angle
   time

Science
Earth science
   astronomy
      moon

Integrated Processes
Observing
Collecting and recording data
Organizing data
Interpreting data
Inferring
Generalizing
Applying

Materials
For each pair of students:
   card stock
   string, 20 cm
   paper clip
   ruler
   transparent tape

For each student:
   colored pencils in two colors
   student pages

Background Information

The moon orbits around the Earth in a cycle that takes just over 27 days to complete. The rotation of the Earth gives the appearance that the moon is moving across the sky each night (and/or day). From one day to the next, the moon rises approximately 50 minutes later than the day before.

When students graph the elevation of the moon, the appearance of the resulting graph will depend on the times of the observations. If all observations are made before the moon reaches its zenith (the highest point in the sky) for the day, the data points will go up in a linear fashion. If observations are taken on both sides of the moon's zenith for the day, the graph will go up, peak, and go back down. If observations are made after the moon's zenith for the day, the graph of the data will go down in a linear fashion. Depending on which kind of data they collect, students should be able to use their graphs to predict either the moonrise, the zenith, or the moonset for the third day.

Times of moonrise and moonset depend on a variety of factors, including the phase of the moon, your location on Earth, and the time of year. However, some generalizations can be made to help you select an appropriate time for doing this activity with your students. The full moon rises around sunset, so if you are able to do this activity as an evening exploration, the days leading up to the full moon would be ideal for observation. The first quarter moon rises around noon, so the days immediately prior to the first quarter moon will allow for four or five hours of moon observation during the school day. The third quarter moon sets around noon, so the days immediately following that will also allow for four of five hours of daylight observation during school hours.

Management

1. This activity requires two consecutive days of moon observation. Make sure that the moon will be visible in the sky at the same times on both days. To ensure this, start your observations on the first day at least two hours after the moon has risen. See *Background Information* for general guidelines, and check your newspaper or the Internet to find local moonrise and moonset times. (See *Internet Connections* for websites with this information.)
2. Select an open outdoor location for making observations where the view of the sky will not be obstructed by trees, buildings, etc. Students should try to be in approximately the same place for each reading.
3. Each pair of students will need one luneometer. Each luneometer requires a copy of the construction page on card stock, a ruler, a 20-cm length of string, a paper clip, and transparent tape.
4. Construct a luneometer ahead of time to use as an example.

Procedure

Day One

1. Ask the *Key Question* and state the *Learning Goals*. Have students share their ideas about how they might track the movement of the moon.
2. Tell them that there is a way for them to record the angle of the moon in the sky, thereby tracking its movement.
3. Show the class a completed luneometer and demonstrate how it works. Distribute the materials for one luneometer to every two students, and have them work together to construct it.
4. When everyone has completed their luneometers, take the class outside and practice using the luneometers by finding the angle of the sightings of the tops of buildings, trees, etc.
5. Distribute the first student page and take students to the observation area. Locate the moon in the sky. Have students record the current time and have partners work together to determine the moon's angle of elevation.
6. Take elevation measurements every 30 minutes for the next two or three hours, as time allows. Being consistent with the observation times helps students easily recognize the pattern.
7. After taking the final reading for the day, distribute the second student page. Have students use their data to graph the elevation of the moon at the different times they observed it.

Day Two

1. Go outside at the same times and repeat the observations from the day before. Match the times from day to day as closely as possible.
2. After all the data are collected, have students graph the data from the second day using a different color than they did the day before.
3. Distribute the final student page and have students respond to the questions. For the final question, the kind of data you collected will determine if they should be able to predict the moonrise, the zenith, or the moonset.
4. Close with a time of discussion where students share their thinking and observations with the class.

Connecting Learning

1. What was your first elevation reading on the first day?
2. How does this compare to the readings of other groups?
3. What might account for some of the differences?
4. What does your graph of your elevation data from the first day look like? What does this tell you?
5. If your readings had been more accurate, do you think the graph would have looked any different? How?

6. What was your first elevation reading on the second day? How does this compare to the first day's reading at the same time?
7. Were the differences between the data from day to day consistent? What do the differences tell you about how the moon moves?
8. What time do you predict the moon will rise (set, reach its zenith) tomorrow? Why?
9. What are you wondering now?

Internet Connections
Sunrise, Sunset Calendars
http://www.sunrisesunset.com
Print your own custom sunrise and sunset calendar including moonrise and moonset times and moon phase information. Select from a list of predefined cities, or enter latitude and longitude for information on any location on Earth.

Weather Underground
http://www.wunderground.com
Enter your city and state to get to your local forecast page. Look under the astronomy section for moonrise and moonset times.

* Reprinted with permission from *Principles and Standards for School Mathematics*, 2000 by the National Council of Teachers of Mathematics. All rights reserved.

THE MOON SHINES BRIGHT

0-3081-984

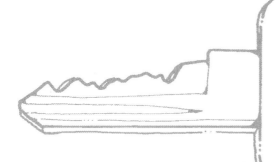

Key Question
What can you learn about the movement of the moon by observing it at different times during the day?

Learning Goals

Students will:

- observe and record the location of the moon in the sky at the same time on two consecutive days, and
- use the information to describe the cycle of the moon and how its location changes from day to day.

THE MOON SHINES BRIGHT
DATA COLLECTION   0010010101101001001010110   0-3081-984

In this investigation you, will chart the movement of the moon for the same two or three hours on two consecutive days.

Carefully record the time and make a measurement of the moon's angle using your luneometer. One person needs to point the luneometer at the moon while the other person reads the number on the scale.

Measure every 30 minutes and record the information in the table. Be sure to take your measurements at the same times on both days.

TIME: ▼	MOON'S ANGLE	
	Day One	Day Two

THE MOON SHINES BRIGHT

Make a line graph of your results from each day. Use a different color pencil to make the lines for each day.

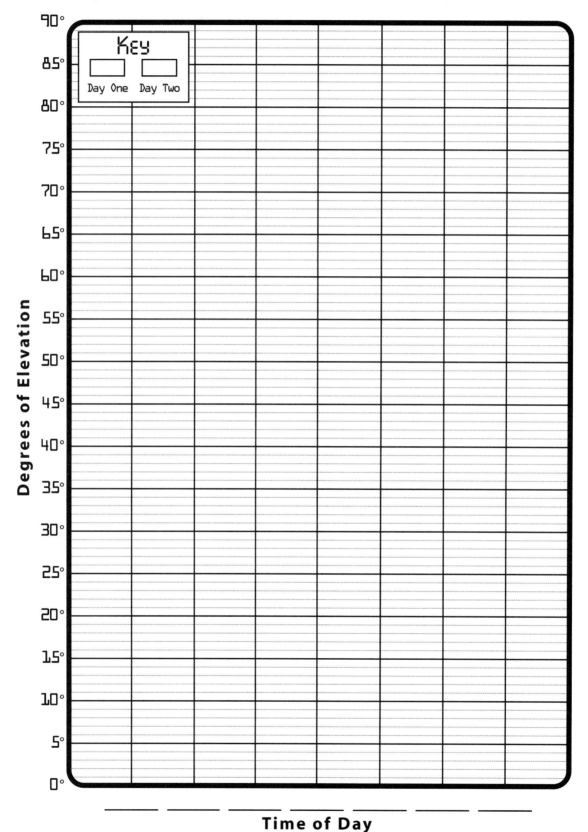

Degrees of Elevation

Key

Day One   Day Two

Time of Day

THE MOON SHINES BRIGHT

DATA ANALYSIS

0-3081-984

1. What does your data tell you about how the moon moves in the sky?

2. Describe the shape of your graph. How does the shape compare from one day to the next?

3. What would you expect the graph of the third day to look like? Why?

4. Use your data to predict either the moonrise, moon set, or the zenith of the moon tomorrow. Describe how you reached your prediction.

59

Constructing Your Luneometer

1. You will need a copy of the lunometer pattern on card stock, a 20 cm piece of string, a paper clip, a ruler, and transparent tape to construct your lunometer.

2. Poke a hole at the spot labeled Plumb Bob Anchor. Thread the string through this hole so that about 2 cm hangs on the back side of the lunometer. Tape this short end securely to the back.

3. Tie a paper clip to the other end of the string and make sure that it swings freely from the vertex.

4. To provide a sighting plane, tape the broad upper band of the lunometer to your ruler.

THE MOON SHINES BRIGHT

0-3081-984

Connecting Learning

1. What was your first elevation reading on the first day?

2. How does this compare to the readings of other groups?

3. What might account for some of the differences?

4. What does your graph of your elevation data from the first day look like? What does this tell you?

5. If your readings had been more accurate, do you think the graph would have looked any different? How?

6. What was your first elevation reading on the second day? How does this compare to the first day's reading at the same time?

Connecting Learning

7. Were the differences between the data from day to day consistent? What do the differences tell you about how the moon moves?

8. What time do you predict the moon will rise (set, reach its zenith) tomorrow? Why?

9. What are you wondering now?

0016

OUT OF THIS WORLD

LINING UP THE PLANETS

0010010101101001001010101

0010010101101001001010101

0-3081-984

Topic
Planet order in the solar system

Key Question
As a team, how can we construct a model of planet order?

Learning Goals
Students will:
- use a collaborative approach to construct a two-dimensional model that represents planet order in the solar system,
- make a scale model that represents planet order, and
- use a reading passage to collect information.

Guiding Documents
Project 2061 Benchmarks
- *Make sketches to aid in explaining procedures or ideas.*
- *In something that consists of many parts, the parts usually influence one another.*
- *A system can include processes as well as things.*

NRC Standard
- *Develop descriptions, explanations, predictions, and models using evidence.*

NCTM Standard 2000*
- *Model problem situations with objects and use representations such as graphs, tables, and equations to draw conclusions*

Math
Logical reasoning

Science
Earth science
  astronomy
    order of planets

Integrated Processes
Observing
Sorting and classifying
Collecting and recording data
Interpreting data
Inferring

Materials
For each four-member student group:
  individual student clue cards
  scale drawings of the planets
  construction paper, 12 x 18 inch

Background Information
The order of the planets in our solar system is frequently addressed in elementary Earth Science. The first four planets (Mercury, Venus, Earth, and Mars) are called the inner planets. Jupiter, Saturn, Uranus, and Neptune make up the outer planets. An asteroid belt creates a dividing line between the inner and outer planets.

Management
1. Make one set of clue cards for each group. Copy them on various colors, or mark the back of the cards to keep track of the different sets.
2. Let the students create a two-dimensional model of the order of the planets.
3. Make sure students have an understanding of inner and outer planets.

Procedure
1. Ask the *Key Question* and state the *Learning Goals*.
2. Divide the class into teams of four.
3. Explain that each team member will have a card that he or she reads to the group. Other team members are not to read each other's cards, but they are to practice listening and then apply what they hear.
4. Have the students use the scaled planet drawings to construct a model of planetary order based on the collaborative clue cards.
5. After the students have constructed the model, have them label the inner and outer planets.
6. Discuss with the students the information obtained from their model.

Connecting Learning
1. What are the eight planets that make up our solar system?
2. What is the order of the eight planets?
3. What are the planets that make up the inner planets? ...the outer planets?
4. How do scientists know this information?
5. How successful was your team in solving the problem? Were some clues more important than others? Explain.
6. Why do you think scientists divide the planets into these two groups?
7. What are you wondering now?

* Reprinted with permission from *Principles and Standards for School Mathematics*, 2000 by the National Council of Teachers of Mathematics. All rights reserved.

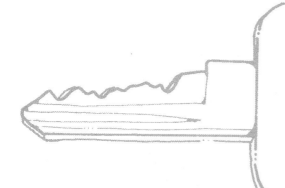

Key Question

As a team, how can we construct a model of planet order?

Learning Goals

Students will:

- use a collaborative approach to construct a two-dimensional model that represents planet order in the solar system,
- make a scale model that represents planet order, and
- use a reading passage to collect information.

LINING UP THE PLANETS

0-3081-984

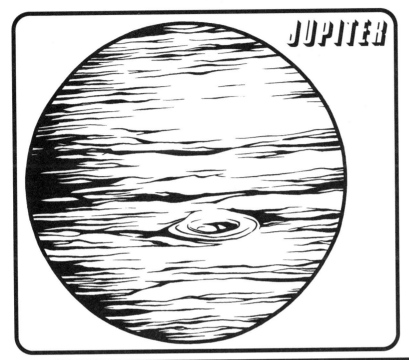

JUPITER

VENUS

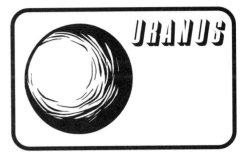

URANUS

MARS

SATURN

NEPTUNE

EARTH

MERCURY

LINING UP THE PLANETS

CLUE CARD 1

- There are eight planets in our solar system.
- Neptune is the farthest planet from the sun.
- Mars is next to Jupiter.
- There are six planets between Mercury and Neptune.

CLUE CARD 2

- Mercury, Earth, Mars, and Venus are called the inner planets.
- Jupiter is the largest planet in the solar system.
- The two largest planets are next to each other.
- There is one planet between Uranus and Jupiter.

CLUE CARD 3

- The outer planets are Saturn, Jupiter, Uranus, and Neptune.
- Mercury and Venus have no moons.
- Earth is the third planet from the sun.
- There is one planet between Venus and Mars.

CLUE CARD 4

- Mercury is the planet closest to the sun.
- Saturn is the second largest planet in the solar system.
- Uranus is between Saturn and Neptune.
- Saturn is closer to Neptune than Earth.

66

LINING UP THE PLANETS

0010010101101001001010
0-3081-984

Connecting Learning

1. What are the eight planets that make up our solar system?

2. What is the order of the eight planets?

3. What are the planets that make up the inner planets? ...the outer planets?

4. How do scientists know this information?

5. How successful was your team in solving the problem? Were some clues more important than others? Explain.

6. Why do you think scientists divide the planets into these two groups?

7. What are you wondering now?

Interesting Facts

Mercury
- its day (176 Earth days) is longer than its year (88 Earth days)

Venus
- has the largest temperature extremes of any planet: 450°C (840°F) on the sunny side to -185°C (-300°F) on the dark side. Some craters at the poles are even colder -212°C (-350°F)
- has no small craters because meteoroids burn up in its dense atmosphere before they can hit its surface

Earth
- is the only planet on which water can exist in the form of a liquid
- 71 percent of its surface is covered with water

Mars
- has largest mountain in solar system: Olympus Mons—24 kilometers (78,000 ft) high with a base diameter of 500 kilometers (310 miles)
- has a system of canyons: Valles Marineris—4000 kilometers (2480 mi) long, 2-7 kilometers (6547-22,915 ft) deep

Jupiter
- twice as massive as all the other planets combined

Saturn
- oblate because it spins so fast that its gases bulge at the middle

Uranus
- many of its moons were named after Shakespeare characters rather than mythological gods

Neptune
- has the fastest winds in the solar system—2000 kilometers per hour (1240 miles per hour)

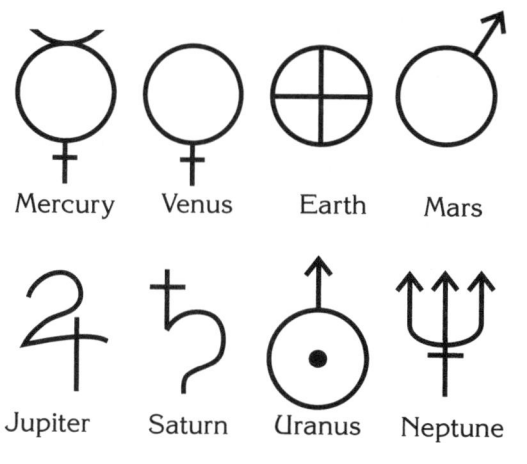

PLANETS IN OUR SOLAR SYSTEM

The word planet comes from the Greek word *planetai,* meaning wanderers. To the ancient peoples, the planets in their orbits seemed to move around the sky. The Greeks and Romans named the planets after mythological gods and goddesses. The symbols used to represent the planets often reflect these names.

Mercury   Venus   Earth   Mars

Jupiter   Saturn   Uranus   Neptune

Dwarf Planets

In 2006, the International Astronomical Union (IAU) decided on a definition of a planet. A planet must:
- orbit the sun,
- be roughly spherical in shape, and
- have cleared other objects out of its orbital path.

Based on this new definition, Pluto, discovered in 1930, is no longer a planet. Instead, it is considered a dwarf planet. Ceres, which is in the asteroid belt between Mars and Jupiter, is also in this category. The number of dwarf planets will continue to grow as astronomers discover more objects in the solar system and learn more about the objects already discovered.

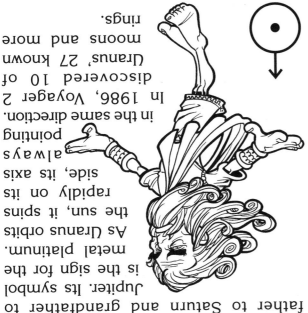

Uranus

Ancient people could not see this seventh planet or any beyond because they did not have telescopes. Following the discovery of the telescope in 1781, astronomers named Uranus for the Roman god who was father to Saturn and grandfather to Jupiter. Its symbol is the sign for the metal platinum. As Uranus orbits the sun, it spins rapidly on its side, its axis always pointing in the same direction. In 1986, Voyager 2 discovered 10 of Uranus' 27 known moons and more rings.

Venus

The thick acid clouds that surround Venus reflect sunlight and we see it as the lovely morning and evening "star." Venus was named for the Roman goddess of love and beauty. Its symbol, a hand mirror, is also the universal symbol for women.

Neptune

A strange pull on the planet Uranus gave astronomers the idea that Neptune existed. Discovered in 1846, the new planet was named for the Roman god of the sea. Its symbol is his fishing spear, the trident. Little is known about the smallest of the gas giants. It has 13 moons and faint rings.

Mercury

Mercury, the planet closest to the sun, is a small, dense, fast-moving planet. It circles the sun in a mere 88 Earth days. The winged helmet and snake-entwined staff of Mercury, the messenger of the Roman gods, can be seen in the symbol for this planet.

Jupiter

Jupiter, the largest planet, is aptly named for the king of the Roman gods and symbolized by his lightning bolt. Jupiter fascinates us with its 63 known moons, swirling poisonous gases, and Great Red Spot. This spot is more than three times the size of the Earth. In it a giant storm has raged for hundreds of years. Voyagers 1 and 2 detected faint rings.

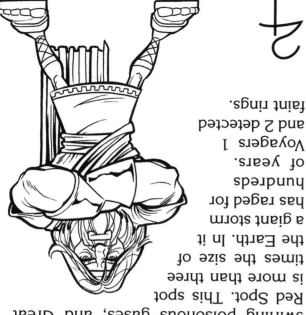

Mars

The reddish mineral covering Mars gives it its nickname, the Red Planet. Because of its color, it was named for the Roman god of war who was covered with blood. His shield and spear form the planet's symbol, which is also the universal sign for man. Mars also tilts on its axis, giving it seasons. Since Mars takes twice as long to circle the sun, its seasons are twice as long as Earth's.

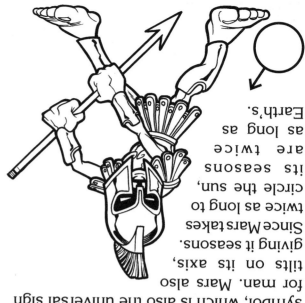

Saturn

Early astronomers described Saturn's rings as "ears." This beautiful planet is a huge, floating ball of gas. Saturn has at least 60 moons. Titan, the largest, is bigger than Mercury and has its own atmosphere. Because of its slow movement, Saturn was named for the Roman god of reaping or time whose symbol is a sickle.

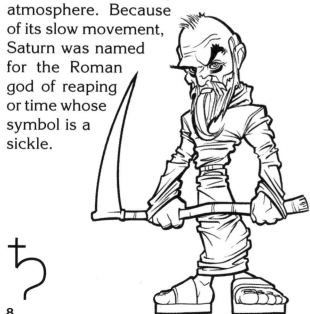

Earth

The planet we know best orbits within the "ecosphere," just the right distance from the sun to provide us with the temperatures and water necessary for our forms of life. The tilt of its axis gives us our seasons. The symbol for Earth is the Greek sign for sphere. Earth is the only planet not named for an ancient god.

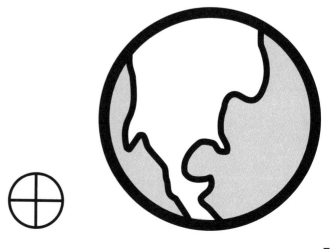

CAN YOU PLANET?

0001

Topic
Planets

Key Question
How can we classify the eight planets?

Learning Goals
Students will:
- read tables and charts to learn about various aspects of the planets and their relationships with one another, and
- use Venn diagrams and graphs to organize information about the planets.

Guiding Documents
Project 2061 Benchmarks
- *The earth is one of several planets that orbit the sun, and the moon orbits around the earth.*
- *Like all planets and stars, the earth is approximately spherical in shape. The rotation of the earth on its axis every 24 hours produces the night-and-day cycle. To people on earth, this turning of the planet makes it seem as though the sun, moon, planets, and stars are orbiting the earth once a day.*
- *Tables and graphs can show how values of one quantity are related to values of another.*
- *Graphical display of numbers may make it possible to spot patterns that are not otherwise obvious, such as comparative size and trends.*
- *Use numerical data in describing and comparing objects and events.*

NRC Standards
- *Mathematics is important in all aspects of scientific inquiry.*
- *The earth is the third planet from the sun in a system that includes the moon, the sun, eight other planets and their moons, and smaller objects, such as asteroids and comets. The sun, an average star, is the central and largest body in the solar system.*
- *Most objects in the solar system are in regular and predictable motion. Those motions explain such phenomena as the day, the year, phases of the moon, and eclipses.*

NCTM Standard 2000*
- *Represent data using tables and graphs such as line plots, bar graphs, and line graphs*

Math
Using Venn diagrams
Graphing
Equalities and inequalities
Whole number operations

Science
Earth science
   astronomy
      planets

Integrated Processes
Observing
Comparing and contrasting
Classifying
Recording data
Interpreting data
Drawing conclusions

Materials
Student pages
Crayons or colored pencils
Scissors

Background Information
Much has been discovered about our planets as a result of information gathered by Voyagers 1 and 2. Students should be encouraged to look for articles that continue to report on new information about our solar system. An excellent web site for current information is http://nineplanets.org.

Management
1. Divide the class into pairs or cooperative learning groups for this activity. Alternate between small group activity and whole group discussions.
2. Each group of students needs one set of planet symbols to manipulate on the Venn diagrams. There are seven sets of symbols on the page.
3. The two pages with pictures of the planets and the sun can be copied and glue together as a visual representation of the relative sizes of the planets.

Procedure
1. Discuss with students what they already know about the planets. (total number [eight], appearance, distance from the Earth, etc.). Have them tell their sources of information whenever possible.

2. Discuss the *Key Question:* How can we classify the eight planets? [size, appearance, having moons, etc.]
3. Distribute the first two student pages. Have students use the *Planetary Facts* information to compare the size of the planets, whether they have rings, and whether they have moons.
4. Guide the students to choose three more attributes with which to classify the planets. Have groups compare their results and discuss any differences.
5. Distribute the planet symbols and the Venn diagram pages. Have the students cut apart one set of symbols and use them to complete the Venn diagrams. Once they have correctly arranged the symbols, have them record the names of the planets in the correct spaces.
6. As a whole class, discuss similarities and differences of the planets from information recorded on the Venn diagrams.
7. Hand out the two graph pages and instruct students to use the information in the *Planetary Facts* table to complete the graphs.
8. Distribute the final page of questions and allow time for students to complete the page.
9. With the whole class, make a list of what has been learned.

Connecting Learning
1. Which planets have rings? [Jupiter, Saturn, Uranus, Neptune] What fraction of the planets is that? [1/2]
2. Which planets have moons? [Earth, Mars, Jupiter, Saturn, Uranus, Neptune] What fraction of the planets is that? [3/4]
3. Which planets are larger than the Earth? [Jupiter, Saturn, Uranus, Neptune] What fraction of the planets is that? [1/2]
4. Which planets have days longer than 24 hours? [Mercury, Venus, Mars] What fraction of the planets is that? [3/8]
5. Would you like to live on a planet with a longer day? How do you think your life would change?
6. How many planets were there in the intersection of the three-circle Venn diagram? [four] Which planets were they? [Jupiter, Saturn, Uranus, Neptune]

7. Do more planets have rings or days longer than 24 hours? [more planets have rings] How do you know?
8. Which planet has the most known moons? [Jupiter] Where did you find this information?
9. What is the total number of known moons in our solar system? [166]
10. Which two planets are the closest in size? [Venus and Earth]
11. Which two planets have the closest length of day? [Earth and Mars] How many minutes difference is there between them? [41 minutes]
12. Which data display did you find to be the most helpful? Why?
13. What are you wondering now?

Extensions
1. Enlarge the Venn diagrams so that they will accommodate the cutouts of the planets. Arrange the planets by a variety of attributes such as
   • smallest to largest
   • longest day to shortest day
   • no moons to most moons
   Be sure students label each continuum clearly: which is smallest, etc.
2. Research information on newly-discovered planet-like objects such as Sedna and Quaoar.

Curriculum Correlation
Language Arts
Have students do research reports on individual planets. The *National Geographic* is an excellent source.

Art
Let each group choose a planet to make in papier-mâché by covering a balloon. Have students research the visual characteristics of their planet to represent it as accurately as possible without regard to its size in relation to other planets. Challenge students to create unique ways to show features such as the rings!

* Reprinted with permission from *Principles and Standards for School Mathematics,* 2000 by the National Council of Teachers of Mathematics. All rights reserved.

CAN YOU PLANET?

10-0820-02

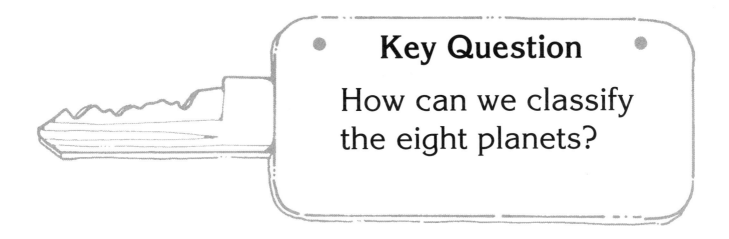

Key Question

How can we classify the eight planets?

Learning Goals

Students will:

- read tables and charts to learn about various aspects of the planets and their relationships with one another, and

- use Venn diagrams and graphs to organize information about the planets.

CAN YOU PLANET?

PLANETARY FACTS

PLANET ▼	Approximate Diameter	Approximate Period of Rotation	Moons	Rings?	
Mercury	4900 km	59 days (176 days) *	0	No	
Venus	12,100 km	243 days (117 days) *	0	No	
Earth	12,800 km	23 hours, 56 minutes	1	No	
Mars	6,800 km	24 hours, 37 minutes	2	No	
Jupiter	143,000 km	9 hours, 55 minutes	63	Yes	
Saturn	120,600 km	10 hours, 39 minutes	60**	Yes	
Uranus	51,100 km	17 hours, 14 minutes	27	Yes	
Neptune	49,500 km	16 hours, 7 minutes	13	Yes	

*   length of day sunrise to sunrise
** new moons are constantly being discovered

Sort out the planets. Next to each planet's name, color in those spaces that are true.

	Larger than Earth	Has Ring(s)	Has Moon(s)
Mercury			
Venus			
Earth			
Mars			
Jupiter			
Saturn			
Uranus			
Neptune			

Select three different characteristics and compare the planets according to those.

Mercury			
Venus			
Earth			
Mars			
Jupiter			
Saturn			
Uranus			
Neptune			

CAN YOU PLANET?

Cut out these symbols to use on the Venn Diagrams.

Mercury	Mercury	Mercury	Mercury	Mercury	Mercury	Mercury
Venus	Venus	Venus	Venus	Venus	Venus	Venus
Earth	Earth	Earth	Earth	Earth	Earth	Earth
Mars	Mars	Mars	Mars	Mars	Mars	Mars
Jupiter	Jupiter	Jupiter	Jupiter	Jupiter	Jupiter	Jupiter
Saturn	Saturn	Saturn	Saturn	Saturn	Saturn	Saturn
Uranus	Uranus	Uranus	Uranus	Uranus	Uranus	Uranus
Neptune	Neptune	Neptune	Neptune	Neptune	Neptune	Neptune

76

Use the information from the chart to place the planets in the correct circle or intersection of circles.

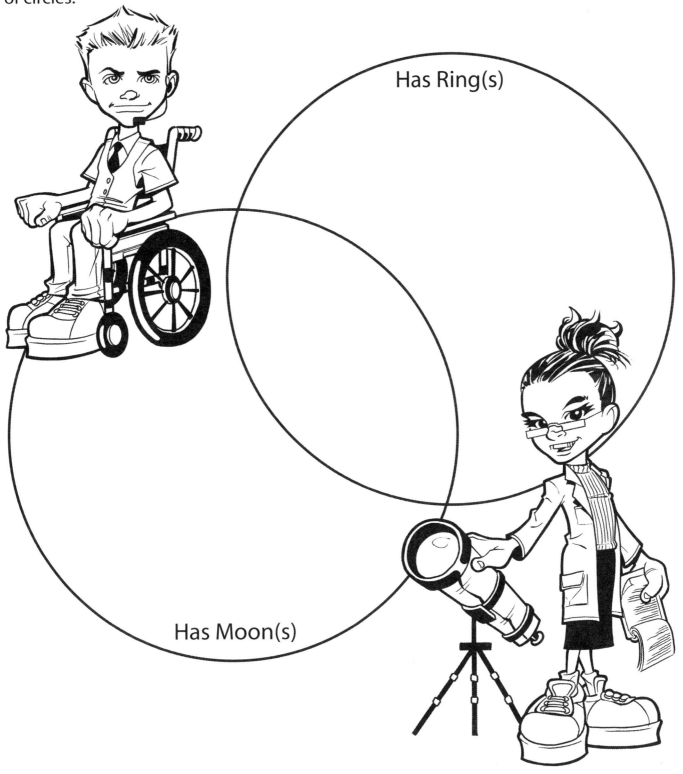

Has Ring(s)

Has Moon(s)

77

CAN YOU PLANET?

Use the information from the chart to place the planets in the correct circle or intersection of circles.

Larger than Earth

Has Moon(s)

Has Ring(s)

78

Graph the number of moons each planet has.

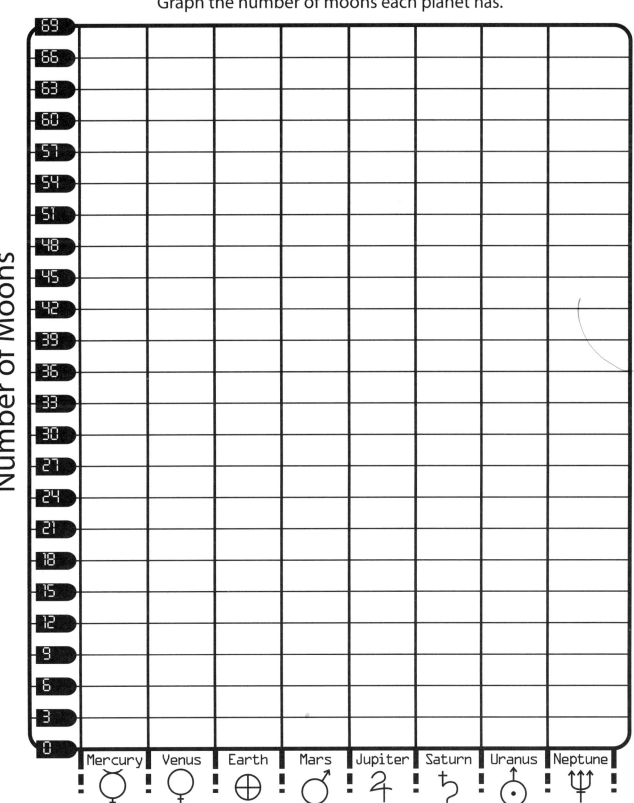

0001

OUT OF THIS WORLD

CAN YOU PLANET?

WHAT'S MY SIZE?

0082-02

Graph the diameters of the planets.

Diameters of Planets

Kilometers ▶	160,000	150,000	140,000	130,000	120,000	110,000	100,000	90,000	80,000	70,000	60,000	50,000	40,000	30,000	20,000	10,000	0
Mercury																	
Venus																	
Earth																	
Mars																	
Jupiter																	
Saturn																	
Uranus																	
Neptune																	

CAN YOU PLANET?

Use the Venn Diagrams or charts to answer the following questions.

1. Which planets are larger than Earth?

2. Which two planets are closest in size?

3. What percent of the planets are smaller than Earth?

4. Which planet have moons?

5. Which planet has the most moons?

6. What is the total number of known moons in our solar system?

7. What is the average number of moons per planet?

8. Which planets have days that are longer than 24 hours?

Think of two more questions you can ask your classmates. Write them below.

THE SOLAR SYSTEM

MERCURY
VENUS
EARTH
MARS

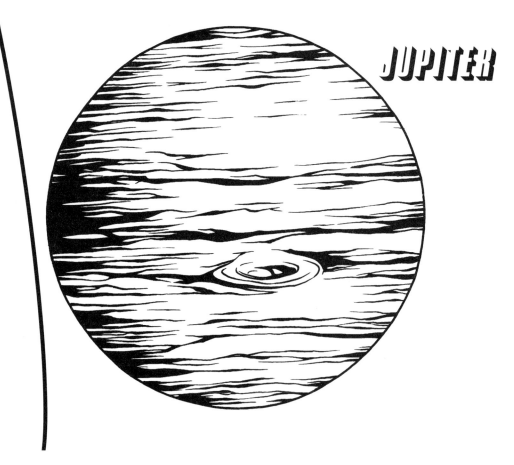

JUPITER

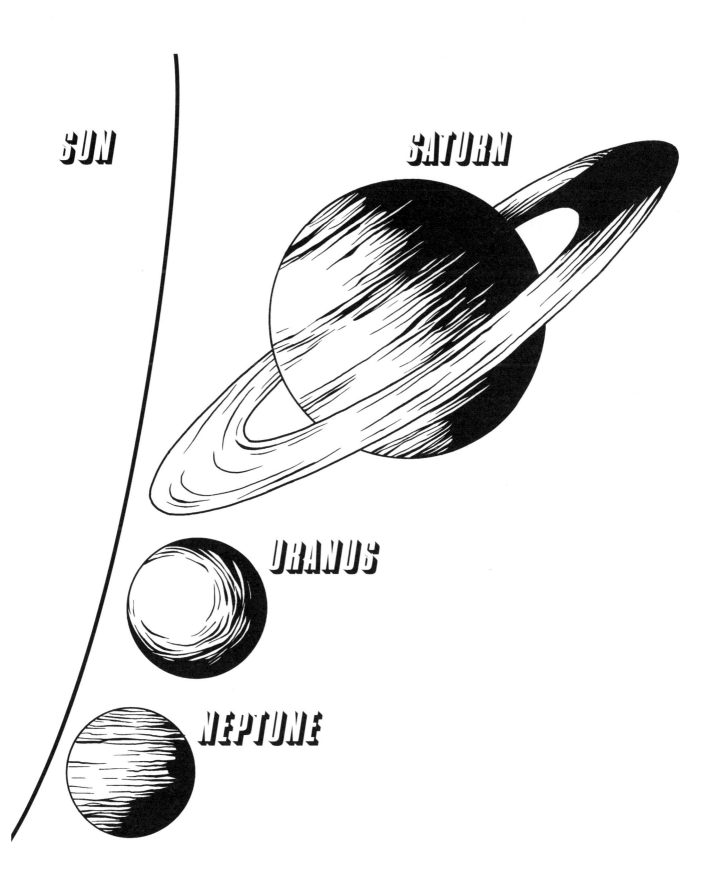

SUN

SATURN

URANUS

NEPTUNE

CAN YOU PLANET?

Connecting Learning

1. Which planets have rings? What fraction of the planets is that?

2. Which planets have moons? What fraction of the planets is that?

3. Which planets are larger than the Earth? What fraction of the planets is that?

4. Which planets have days longer than 24 hours? What fraction of the planets is that?

5. Would you like to live on a planet with a longer day? How do you think your life would change?

6. How many planets were there in the intersection of the three-circle Venn diagram? Which planets were they?

7. Do more planets have rings or days longer than 24 hours? How do you know?

Connecting Learning

8. Which planet has the most known moons? Where did you find this information?

9. What is the total number of known moons in our solar system?

10. Which two planets are the closest in size?

11. Which two planets have the closest length of day? How many minutes difference is there between them?

12. Which data display did you find to be the most helpful? Why?

13. What are you wondering now?

Classifications of Planets

Their Composition

Terrestrial Planets: Mercury, Venus, Earth, Mars

    These planets are made up mostly of rock and metal.

    They have higher densities and very few moons.

    They rotate relatively slowly.

Gas Planets: Jupiter, Saturn, Uranus, Neptune

    These planets are made up mostly of hydrogen and helium gas.

    These planets have lower densities and lots of moons.

    They rotate relatively rapidly.

Their Size

Small Planets: Mercury, Venus, Earth, and Mars

    diameters <13,000 km

Giant Planets: Jupiter, Saturn, Uranus, and Neptune

    diameters >48,000 km

    (These planets are sometimes called the gas giants.)

Their Location from the Sun

Inner Planets: Mercury, Venus, Earth, and Mars

Outer Planets: Jupiter, Saturn, Uranus, and Neptune

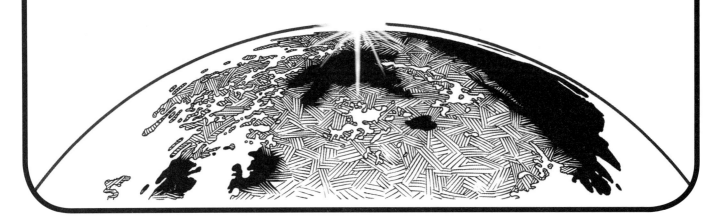

PLANETARY FACTS

Topic
Planets

Key Question
What can we learn about the planets by displaying the information in Venn diagrams?

Learning Goals
Students will:
- gather information about the eight planets,
- create tables and Venn diagrams to display the information, and
- interpret the information.

Guiding Documents
Project 2061 Benchmark
- *The earth is one of several planets that orbit the sun, and the moon orbits around the earth.*

NRC Standard
- *The earth is the third planet from the sun in a system that includes the moon, the sun, eight other planets and their moons, and smaller objects, such as asteroids and comets. The sun, an average star, is the central and largest body in the solar system.*

Math
Using Venn diagrams

Science
Earth science
astronomy
planets

Integrated Processes
Observing
Comparing and contrasting
Classifying
Recording data
Interpreting data

Materials
For the class:
research materials or tables of information

Background Information
Venn diagrams are used to represent sets with something in common. They can be used to compare and contrast two or more attributes. If students have not previously used Venn diagrams, begin with one circle. The inside of this circle is for planets that have a certain attribute and the area surrounding the circle is for planets that do not have that attribute.

A two-circle Venn has three areas plus the area outside the circles. The intersection of the two circles is for planets with both attributes. Again, the area outside the circles is for planets that do not have either attribute.

The three-circle Venn diagram follows the same pattern. Now, there are seven regions. Each region overlaps each of the others, and the center is an overlap of all three attributes. Planets within this area will have all three attributes.

Attributes for this lesson can be obtained from the data tables within the lesson or from students' own research.

Management
1. Time for this activity will vary depending on whether students use the table that is provided or need to do research to gather data.
2. Students should work in small groups to gather data, create tables, and generate the Venn diagram.

Procedure
1. Ask the *Key Question* and state the *Learning Goals*.
2. If necessary, review procedures for making and using Venn diagrams.
3. Tell students they need to research three topic areas about the eight planets.
4. Direct them to make an organized chart of the information.
5. Have them display the topic areas on the Venn diagram and determine the area in which the planets should be displayed.
6. Invite the students to generate discussion questions that pertain to their Venn diagrams.
7. Direct groups to exchange questions and Venn diagrams. Have the groups check the answers of their partnering groups.
8. Encourage students to share with the class the information they found out.

Connecting Learning

1. Which attributes did you chose to display in the Venn diagram?
2. Were there some attributes that were easier to use in a Venn diagram than others? Explain.
3. What kind of questions can you ask about each of the Venn diagrams?
4. What value do you find in putting your information into a Venn diagram? Explain.
5. What are you wondering now?

Internet Connections

The Eight Planets
http://nineplanets.com
This site gives a variety of information—temperature, diameter, mass, number of moons, planetary composition, etc. Difficult reading level, but students can glean interesting information.

Enchanted Learning
http://www.enchantedlearning.com/subjects/astronomy/planets/
Many graphs are provided to convey information about the planets. Appropriate reading level.

Journey Through the Galaxy
http://filer.case.edu/~sjr16/advanced/planets_main.html
Includes a large data table with information students can use for making the Venn diagrams.

NASA
http://nssdc.gsfc.nasa.gov/planetary/factsheet/
This site has a large data table in metric units and one in U.S. units.

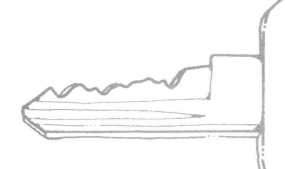

Key Question

What can we learn about the planets by displaying the information in Venn diagrams?

Learning Goals

Students will:

- gather information about the eight planets,
- create tables and Venn diagrams to display the information, and
- interpret the information.

PLANETARY FACTS TERMS

Information about our solar system is constantly changing. New information is being obtained through photographs and other data.

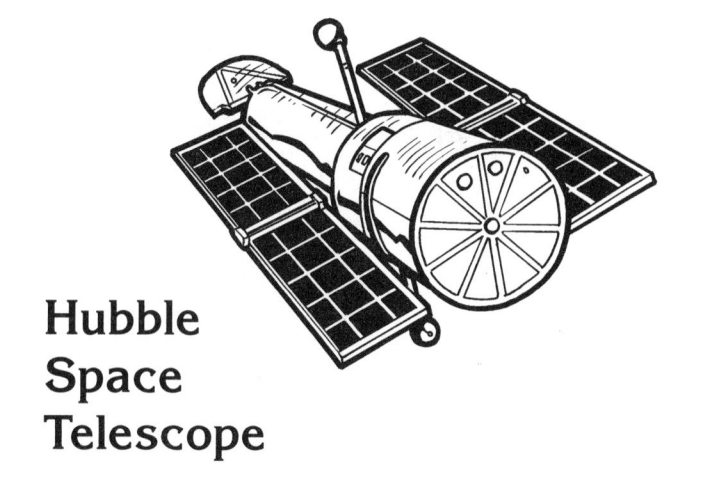

Hubble Space Telescope

SURFACE GRAVITY

Gravity is the force that pulls or holds an object to a planet. The more mass a planet has, the greater its pull and the more you would weigh there. You would weigh just over two and one half times greater on Jupiter than you do on Earth.

JUPITER: 250 LBS

$$\frac{M}{V}$$

DENSITY

Density tells how tightly mass is packed and is determined by the mass of an object divided by its volume. A basket of feathers has the same volume as a basket of lead but the lead is denser, and so it would have a greater mass.

MOONS

This is the number of natural satellites a planet has. Some planets have no moons and some probably have moons we have yet to discover.

DIAMETER This is the length of a line passed through the middle of a sphere. The larger the diameter, the larger the planet or sphere.

The inner planets have a density greater than that of water and would sink in a giant ocean of water. The outer planets are more gaseous and so less dense. Saturn has a density less than water; it would float.

DISTANCE FROM THE SUN AND EARTH

These are measured in millions of kilometers. The distances given are the average distances a planet is orbiting from the sun or the Earth. As the orbits of the planet are elliptical rather than circular, a planet will sometimes be closer and sometimes farther in the course of the year.

LENGTH OF YEAR

This is the amount of time it takes a planet to make one complete orbit around the sun.

EARTH:
1 YEAR = 365 DAYS.
IT'S CONFIRMED.

MASS

This is the amount of material that something contains. Mass and weight are not the same things. An object acquires weight due to the pull of gravity. A person or object is weightless in space but still has mass. For comparison, the mass of the Earth is considered as 1 and other planets' masses are in relation to that.

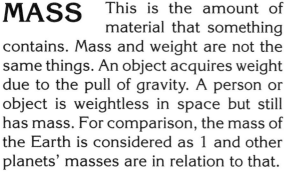

HEY!
I MAY JUST
BE FLOATING HERE,
BUT I'M NO
PUSHOVER!

LENGTH OF DAY

Except for Mercury and Venus, this is very close to how long a day is based on Earth's time.

OUT OF THIS WORLD

PLANETARY FACTS

OUT OF THIS WORLD

0010010101001010010
0010010101001010010
0010010101001010010

0-1201-977

	Distance from Earth in Millions of Km	Distance from Sun in Millions of Km	Density	Mass	Temperature	Gravity	Length of Year
Mercury	91.7	57.9	5.5	0.055	-185° to 450°C	0.39	88 days
Venus	41.4	108.2	5.2	0.815	482°C surface	0.91	225 days
Earth	0	149.6	5.5	1.0	15°C avg. surface	1	365 days
Mars	78.3	227.9	3.9	0.11	-23°C avg. surface	0.38	687 days
Jupiter	628.7	778.3	1.3	318	-150°C at cloud tops	2.53	11.9 years
Saturn	1277	1427	0.7	95.2	-180°C at cloud tops	1.07	29.5 years
Uranus	2721	2870	1.3	15	-210°C at cloud tops	0.91	84 years
Neptune	4347	4497	1.6	17	-220°C at cloud tops	1.15	165 years

PLANETARY FACTS

Make your own Venn diagram using facts about the planets.

94

Connecting Learning

CONNECTING LEARNING

1. Which attributes did you chose to display in the Venn diagram?

2. Were there some attributes that were easier to use in a Venn diagram than others? Explain.

3. What kind of questions can you ask about each of the Venn diagrams?

4. What value do you find in putting your information into a Venn diagram? Explain.

5. What are you wondering now?

SPACING OUT THE SYSTEM

Topic
Planets, relative distances

Key Question
Using the relative distances of the planets, how can we make a model solar system in the class or on the playground?

Learning Goals
Students will:
- determine the relative distances of the planets, and
- use this information to construct two different models of the solar system.

Guiding Documents
Project 2061 Benchmarks
- *Nine planets of very different size, composition, and surface features move around the sun in nearly circular orbits. Some planets have a great variety of moons and even flat rings of rock and ice particles orbiting around them. Some of these planets and moons show evidence of geological activity. The earth is orbited by one moon, many artificial satellites, and debris.*
- *The earth is one of several planets that orbit the sun, and the moon orbits around the earth.*
- *Different models can be used to represent the same thing. What kind of a model to use and how complex it should be depends on its purpose. The usefulness of a model may be limited if it is too simple or if it is needlessly complicated. Choosing a useful model is one of the instances in which intuition and creativity come into play in science, mathematics, and engineering.*

NRC Standard
- *The earth is the third planet from the sun in a system that includes the moon, the sun, eight other planets and their moons, and smaller objects, such as asteroids and comets. The sun, an average star, is the central and largest body in the solar system.*

NCTM Standards 2000*
- *Understand and use ratios and proportions to represent quantitative relationships*
- *Solve problems involving scale factors, using ratio and proportion*

- *Solve problems that arise in mathematics and in other contexts*

Math
Computation
Estimation
  rounding
Measurement
  length
Problem solving

Science
Earth science
  astronomy
    planets

Integrated Processes
Observing
Comparing and contrasting
Recording data
Interpreting data
Applying

Materials
For the class:
  pencils
  calculators
  metric rulers and tapes
  trundle wheel
  12" x 18" paper
  Optional: seeds and nuts (see *Management 4*)

Background Information
Astronomers use astronomical units to measure distances that are too small to be measured in light years. The distance of each planet from the sun is provided on the chart for the activity *Planetary Facts?* Students can refer to that chart or the teacher can provide it. To find the astronomical unit, students use a comparative ratio. By simply dividing the distances in millions of kilometers, students can work with smaller numbers. After they have determined relative distances, they can let the Earth equal any distance they wish and simply multiply that by the relative distance to create their model. Distances given are the average mean distance from a planet to the Earth or the sun. Using the distance of Earth from the sun as one centimeter,

students can space the planets on a piece of 12 x 18 inch paper. Using 0.5 meter to represent Earth's distance from the sun, students can measure relative distances on most school playgrounds.

Relative Distances

Planet	Paper Scale (cm)	Playground Scale (m)
Mercury	.4	.2
Venus	.7	.4
Earth	1.0	.5
Mars	1.5	.8
Jupiter	5.2	2.6
Saturn	9.5	4.8
Uranus	19.1	9.6
Neptune	30.0	15.0

Management

1. This activity will probably take two class periods; one period to complete the computation and another to actually measure the distances.
2. Students can work in pairs or small groups to do this activity.
3. Students need to have some idea of how a ratio works before completing the activity.
4. For plotting the solar system on 12 x 18 inch paper, students can cut circles to represent the planets or use the following seeds and nuts. Radish seeds—Mercury; split pea—Mars; dried peas—Earth and Venus; lima beans—Uranus and Neptune; and walnuts—Jupiter and Saturn.
5. Relative size can be determined using the activity sheet in the following lesson, *Size it Up*.
6. For a playground model, let a building or wall represent the location of the sun. The distance of the Earth may be any amount you decide and depends on the size of your playground. One way to determine this is to have students use metric tapes or a trundle wheel to measure the farthest distance on the playground. From that measurement, they can decide what value they can give to Earth so all the planets will fit in.

Procedure

1. Distribute the first two information pages and go over the content as a class. Ask the *Key Question* and state the *Learning Goals*.
2. Put a transparency of the chart from *Planetary Facts* on the overhead, or provide students with the distance from the sun information.
3. Hand out the first student page and have students work in pairs or small groups to complete the table. They should divide the distance that each planet is from the sun by 150 to determine the relative distance. Answers should be rounded to the nearest tenth.
4. Distribute the second student page and the materials for the classroom model solar system. Have groups complete the chart and work together to create the model.
5. As a class, work together to create the playground model of the solar system.

Connecting Learning

1. What do you notice about the distances from the sun to the inner planets compared with the distances to the outer planets?
2. Which is greater, the distance from the sun to the Earth, or the distance from the Earth to Jupiter? How do you know?
3. How does the classroom model compare to the playground model?
4. Which is a better representation of the solar system? Why?
5. In what ways are our models like the actual solar system? In what ways are they different?
6. If you were to make one improvement to either model, what would it be?
7. What are you wondering now?

* Reprinted with permission from *Principles and Standards for School Mathematics*, 2000 by the National Council of Teachers of Mathematics. All rights reserved.

SPACING OUT THE SYSTEM

0308-198-1

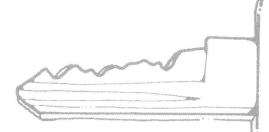

Key Question

Using the relative distances of the planets, how can we make a model solar system in the class or on the playground?

Learning Goals

Students will:

- determine the relative distance of the planets, and
- use this information to construct two different models of the solar system.

SPACING OUT THE SYSTEM
ASTRONOMICAL UNIT

To measure distances in space, astronomers have chosen a unit called the astronomical unit. The length of an astronomical unit is the average distance of the Earth from the sun. The distance is about 93,000,000 miles or 150,000,000 kilometers. The exact figure is not as important to us as the ratio or relative distance of the planets from the sun. Using astronomical units, those relative distances stay the same whether we use miles or kilometers.

Using 150,000,000 kilometers as one astronomical unit, you can create a model solar system in your classroom or on the playground.

The planets travel in elliptical rather than true circular orbits as they make their way around the sun. This means that they are sometimes closer and sometimes farther away. The distances we are using are called the "mean average." This means the average between a planet's farthest point from the sun or Earth and its nearest point. The maximum, minimum, and mean average distances for Earth, Uranus, and Neptune are listed below.

	Earth	Uranus	Neptune
Maximum distance from sun in millions of kilometers	152.1	3004	4537
Minimum distance from sun in millions of kilometers	147.1	2735	4456
Mean distance from sun in millions of kilometers	150	2870	4497

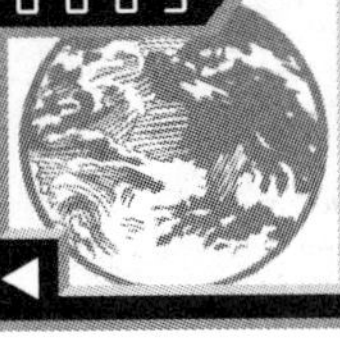

0003 OUT OF THIS WORLD

0010010101101001001010

SPACING OUT THE SYSTEM

0010010101101001001010 0308-198-1

In order to create a model of the solar system that is accurate in terms of distance, you use the astronomical unit, which is the average distance of the Earth from the sun, or 150,000,000 kilometers. The distance of the Earth from the sun is then equal to one unit. By creating a ratio for each of the other planets, you can determine their relative distances. Here is an example using the planet Mercury.

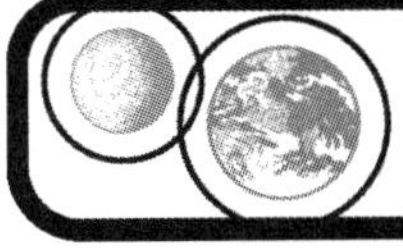

	Distance from the sun in millions of kilometers		Relative Distance
Mercury / Earth	$\frac{58}{150}$	=	$\frac{N}{1}$

Cross multiply and you find that 150 N equals 58. Divide both sides by 150 and you have the relative distance of Mercury from the sun.

Complete the chart below. Round your final answer to the nearest tenth. Round the planet's distances to the nearest million kilometers.

PLANET	Distance in Millions of Kilometers	÷ 150	=	Relative Distance	Rounded to Nearest Tenth
Mercury		÷ 150	=		
Venus		÷ 150	=		
Earth		÷ 150	=		
Mars		÷ 150	=		
Jupiter		÷ 150	=		
Saturn		÷ 150	=		
Uranus		÷ 150	=		
Neptune		÷ 150	=		

SPACING OUT THE SYSTEM

Use the relative distances from the previous page to complete the table below for creating a model solar system for the classroom and playground. Use this information to construct one or both models. Start at one end of a bulletin board or playground and label the sun. Use the chart below and measure to show the relative distances. In the classroom, you can also use objects or circles of relative size. On the playground, have a person hold a sign for each planet and stand the measured distance apart. Take turns observing from a distance.

> ## Classroom: Let Earth equal 1 centimeter
> ## Playground: Let Earth equal .5 meter

PLANET ▼	Classroom	Playground
Mercury		
Venus		
Earth		
Mars		
Jupiter		
Saturn		
Uranus		
Neptune		

102

Connecting Learning

1. What do you notice about the distances from the sun to the inner planets compared with the distances to the outer planets?

2. Which is greater, the distance from the sun to the Earth, or the distance from the Earth to Jupiter? How do you know?

3. How does the classroom model compare to the playground model?

4. Which is a better representation of the solar system? Why?

5. In what ways are our models like the actual solar system? In what ways are they different?

6. If you were to make one improvement to either model, what would it be?

7. What are you wondering now?

SIZE IT UP

Topic
Planets

Key Question
How can we make a model solar system that will show the relative sizes of the planets?

Learning Goals
Students will:
- determine the relative sizes of the planets, and
- use this information to construct a model of the solar system.

Guiding Documents
Project 2061 Benchmarks
- *Nine planets of very different size, composition, and surface features move around the sun in nearly circular orbits. Some planets have a great variety of moons and even flat rings of rock and ice particles orbiting around them. Some of these planets and moons show evidence of geological activity. The earth is orbited by one moon, many artificial satellites, and debris.*
- *The earth is one of several planets that orbit the sun, and the moon orbits around the earth.*
- *Different models can be used to represent the same thing. What kind of a model to use and how complex it should be depends on its purpose. The usefulness of a model may be limited if it is too simple or if it is needlessly complicated. Choosing a useful model is one of the instances in which intuition and creativity come into play in science, mathematics, and engineering.*

NRC Standard
- *The earth is the third planet from the sun in a system that includes the moon, the sun, eight other planets and their moons, and smaller objects, such as asteroids and comets. The sun, an average star, is the central and largest body in the solar system.*

NCTM Standards 2000*
- *Understand and use ratios and proportions to represent quantitative relationships*
- *Solve problems involving scale factors, using ratio and proportion*
- *Solve problems that arise in mathematics and in other contexts*

Math
Proportional reasoning
  scale
  ratios
Estimation
  rounding
Problem solving
Decimals
Geometry

Science
Earth science
  astronomy
    planets

Integrated Processes
Observing
Comparing and contrasting
Recording data
Interpreting data
Inferring
Applying

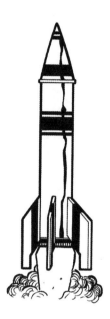

Materials
For the class:
  calculators
  metric rulers and tapes
  construction paper
  pencils
  clay
  drawing compasses

Background Information
  To make a scale model of the solar system, students must give the diameter of the Earth a value such as one centimeter and use comparative ratios to determine the relative sizes of the rest of the planets. The diameter of each planet is provided in the chart for the activity *Can You Planet?* Students can refer to that chart, or the teacher can provide the information. Directions for setting up the ratio are on the student page. After they have determined relative sizes of the planets, the students are to construct clays balls of relative sizes to use in the model solar system. On the second activity page, students are given diameter lines on which to form these balls. They may then construct larger circles from construction paper, using a larger number (such as ten) for the Earth's diameter.

104

Management

1. This activity may take two class periods, one period to complete the computations and another to actually construct the circles.
2. Students can work in pairs or small groups to do this activity.
3. Students need to have some knowledge of ratios before completing the activity.
4. You can use any number to equal the diameter of Earth. One was selected to facilitate the formation of the clay balls.
5. Larger circles can be constructed from paper by determining the radius of each one and using a compass to draw it.
6. In constructing the paper models, students need to divide the diameter by two to find the radius of each planet.

Procedure

1. Ask the *Key Question* and state the *Learning Goals*.
2. Put a transparency of the chart from *Can You Planet?* on the overhead, or provide students with the information necessary.
3. Distribute the first student page and have students work in pairs or small groups to complete the table. They should divide the diameter of each planet by 12,800 to determine the relative diameter. Answers should be rounded to the nearest tenth.
4. Distribute the second student page and the materials for the model solar system. Have groups complete the chart and work together to create the model.

Connecting Learning

1. What do you notice about the sizes of the various planets?
2. Which planet has the largest diameter? ...the smallest diameter?
3. In what ways is our model like the actual solar system? In what ways is it different?
4. If you were to make one improvement to the model, what would it be?
5. What are you wondering now?

* Reprinted with permission from *Principles and Standards for School Mathematics*, 2000 by the National Council of Teachers of Mathematics. All rights reserved.

SIZE IT UP

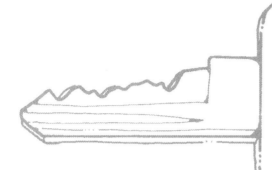

Key Question

How can we make a model solar system that will show the relative sizes of the planets?

Learning Goals

Students will:

- determine the relative sizes of the planets, and
- use this information to construct a model of the solar system.

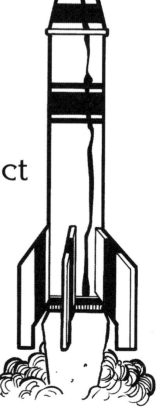

Relative size can be determined in much the same way as relative distance. Give the diameter of Earth a value of 1. By creating a ratio, you can obtain the relative diameter for each planet in the solar system. You can then use that relative diameter to make your own solar system model by setting up a ratio for each of the planets. Here is a sample using Mercury:

Mercury's diameter: $\dfrac{4900}{12{,}800} = \dfrac{N}{1}$
Earth's diameter:

Cross multiply and you get the equation $4900 = 12{,}800\ N$. By dividing both sides by 12,800 you obtain the relative diameter for Mercury.

Complete the chart below for all the planets.

PLANET ▼	Diameter (km) ÷	Earth's Diameter 12,800 =	Relative Diameter
Mercury	÷	12,800 =	
Venus	÷	12,800 =	
Earth	÷	12,800 =	
Mars	÷	12,800 =	
Jupiter	÷	12,800 =	
Saturn	÷	12,800 =	
Uranus	÷	12,800 =	
Neptune	÷	12,800 =	

SIZE IT UP
SOLAR SYSTEM SCALE MODELS

0010010101101001001010010

012-9194-9

⊢ *MERCURY*

⊢—⊣ *VENUS*

EARTH ⊢—⊣

⊢—⊣ *MARS*

1 Make a scale model of the solar system out of clay. The relative diameter for each planet is drawn on the page for you. Form balls of clay with diameters to match. Put the planets in order on your desk.

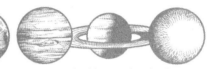

⊢————————————————⊣ *JUPITER*

⊢————————————⊣ *SATURN*

⊢——————⊣ *URANUS*

2 Make a larger model of the solar system from paper. Do the computations then use a compass to construct the circles. Color and post.

NEPTUNE ⊢———⊣

PLANET ▼	Relative Diameter	Scale Diameter (x10)	Radius (÷ 2)
Mercury	.4		
Venus	.9		
Earth	1	10 cm	5 cm
Mars	.5		
Jupiter	11.2		
Saturn	9.4		
Uranus	4.0		
Neptune	3.9		

SIZE IT UP

Connecting Learning

1. What do you notice about the sizes of the various planets?

2. Which planet has the largest diameter? ...the smallest diameter?

3. In what ways is our model like the actual solar system? In what ways is it different?

4. If you were to make one improvement to the model, what would it be?

5. What are you wondering now?

PLANETARY SCAVENGER HUNT

0010010101101001001010
04-0319-54

Topic
Planets

Key Question
What everyday objects can we use to represent the relative sizes of the Earth and the other planets?

Learning Goals
Students will:
- create a model of the solar system using a marble to represent the Earth, and
- find other spherical objects of relative sizes to represent the rest of the planets.

Guiding Documents
Project 2061 Benchmarks
- *Nine planets of very different size, composition, and surface features move around the sun in nearly circular orbits. Some planets have a great variety of moons and even flat rings of rock and ice particles orbiting around them. Some of these planets and moons show evidence of geological activity. The earth is orbited by one moon, many artificial satellites, and debris.*
- *The earth is one of several planets that orbit the sun, and the moon orbits around the earth.*
- *Different models can be used to represent the same thing. What kind of a model to use and how complex it should be depends on its purpose. The usefulness of a model may be limited if it is too simple or if it is needlessly complicated. Choosing a useful model is one of the instances in which intuition and creativity come into play in science, mathematics, and engineering.*

NRC Standard
- *The earth is the third planet from the sun in a system that includes the moon, the sun, eight other planets and their moons, and smaller objects, such as asteroids and comets. The sun, an average star, is the central and largest body in the solar system.*

NCTM Standards 2000*
- *Select and apply appropriate standard units and tools to measure length, area, volume, weight, time, temperature, and the size of angles*
- *Understand that measurements are approximations and how differences in units affect precision*
- *Solve problems that arise in mathematics and in other contexts*

Math
Problem solving
Measurement
Operations
   subtraction

Science
Earth science
   astronomy
      planets

Integrated Processes
Observing
Collecting and recording data
Interpreting data

Materials
For the class:
   pencils
   calculators
   metric rulers and tapes
   balls, marbles, and other spherical objects

Background Information
To make a scale model of the solar system using spherical objects, students must first have an object such as a marble to represent Earth. They then measure the object to determine the constant factor. By multiplying the relative diameters of the other planets by this constant, they determine the diameters their other objects must have. They then locate objects (seeds of various sizes, a variety of marbles, rubber balls, tennis balls, and larger playground balls) to represent the other planets.

Management
1. This activity may be done over an extended period of time as students search for objects. Several objects can be provided in a classroom setting or students can do this as an out-of-class project and share objects they have found.
2. Students should work in pairs or small groups to do this activity.
3. You can use any object to equal the diameter of Earth. A marble was selected because both larger and smaller spheres would be relatively easy to locate.

4. Students determine the diameter of a marble by placing it between two rulers and measuring the distance across. One student holds the two rulers upright against the object while the other student measures the distance between the two rulers.

5. Students need to be reminded that all measured numbers are approximate and that they may not find spheres that are exactly equal to a planet's relative diameter.

Procedure

1. Ask the *Key Question* and state the *Learning Goals*.
2. Distribute a marble and three rulers to each group or pair of students. Demonstrate how to measure the diameter of the marble by placing it between two rulers and measuring how far apart the rulers are.
3. Give each student a copy of the student page. Have them measure their marbles and record the diameters in the *Needed* and *Actual* columns. (The difference for Earth will be zero.)
4. Show them how to use the actual measured diameter of their Earth to complete the *Needed Diameter* column.
5. Give students time to go on a "scavenger hunt" to find other spherical objects that are approximately the correct relative sizes to represent the other eight planets.
6. Once they have measured and recorded all of the actual diameters, have them compute the difference between each planet's actual and needed diameter and record it in the table.
7. Have them compute their scores by adding up all their differences. The group with the lowest score wins.

Connecting Learning

1. What objects did you find to represent the planets?
2. How did the objects you chose compare to the objects other groups chose?
3. Which group had the lowest score? How does their score compare to the group that had the highest score?
4. What does this model show you about the solar system? [the relative sizes of the planets] What does it not show you? [the distance the planets are from the sun and each other, the planets' orbits, etc.]
5. What are you wondering now?

* Reprinted with permission from *Principles and Standards for School Mathematics*, 2000 by the National Council of Teachers of Mathematics. All rights reserved.

PLANETARY SCAVENGER HUNT

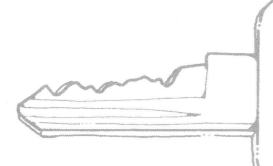

Key Question

What everyday objects can we use to represent the relative sizes of the Earth and the other planets?

Learning Goals

Students will:

- create a model of the solar system using a marble to represent the Earth, and
- find other spherical objects of relative sizes to represent the rest of the planets.

112

PLANETARY SCAVENGER HUNT

Use a marble to represent Earth. Measure its diameter and record. That measurement is the constant factor. Multiply each planet's relative diameter by the constant to determine the needed diameter. Scavenge for spheres with appropriate diameters to represent the planets. Measure, record, and find the difference.

PLANET	Diameter in Km	Relative Diameter	Objects	Needed Diameter	Actual Diameter	Difference
Mercury	4900	.4				
Venus	12,100	.9				
Earth	12,800	1	Marble			
Mars	6800	.5				
Jupiter	143,000	11.2				
Saturn	120,600	9.4				
Uranus	51,100	4.0				
Neptune	49,500	3.9				

SCORE ▶

PLANETARY SCAVENGER HUNT

Connecting Learning

1. What objects did you find to represent the planets?

2. How did the objects you chose compare to the objects other groups chose?

3. Which group had the lowest score? How does their score compare to the group that had the highest score?

4. What does this model show you about the solar system? What does it not show you?

5. What are you wondering now?

EXTRA-TERRESTRIAL EXCURSIONS

Topic
Computation

Key Question
How old would you be when you got to each planet in our solar system if you left Earth today, traveling at a speed of 40,000 kilometers per hour?

Learning Goals
Students will:
- determine the amount of time it would take to get to the moon and each planet in our solar system at a speed of 40,000 kilometers per hour, and
- use this information to determine how old they would be if they were able to travel to the planets at that speed.

Guiding Documents
Project 2061 Benchmarks
- *Add, subtract, multiply, and divide whole numbers mentally, on paper, and with a calculator.*
- *Nine planets of very different size, composition, and surface features move around the sun in nearly circular orbits. Some planets have a great variety of moons and even flat rings of rock and ice particles orbiting around them. Some of these planets and moons show evidence of geological activity. The earth is orbited by one moon, many artificial satellites, and debris.*
- *The earth is one of several planets that orbit the sun, and the moon orbits around the earth.*

NRC Standards
- *The earth is the third planet from the sun in a system that includes the moon, the sun, eight other planets and their moons, and smaller objects, such as asteroids and comets. The sun, an average star, is the central and largest body in the solar system.*
- *Use mathematics in all aspects of scientific inquiry.*

NCTM Standards 2000*
- *Select appropriate methods and tools for computing with whole numbers from among mental computation, estimation, calculators, and paper and pencil according to the context and nature of the computation and use the selected method or tools*
- *Solve problems that arise in mathematics and in other contexts*

Math
Computation
Problem solving
    logic

Science
Earth science
    astronomy
        distances to planets

Integrated Processes
Observing
Collecting and recording data
Interpreting data
Applying

Materials
For the class:
    calculators

Background Information
The distance to the moon and each planet (in kilometers) is listed on the student page. (If the teacher prefers, the students can research these distances.) The speed of 40,000 kilometers per hour was determined by the speed of our space travel. Obviously there would be more to interplanetary travel than how long it would take, and students may bring up all those problems.

Since the answers to the first three computations should be rounded, rounding strategies should be taught prior to this activity. The last computation (converting months to years) should be done without a calculator, as the remainder will tell students the number of months. This serves as a good example of a situation in which computing with a calculator won't work. The final activity sheet, *Special Delivery*, gives students the opportunity to use the information in a problem-solving format.

Management
1. Time for this activity will vary depending on the students' math abilities and whether or not calculators are used.
2. Students can work in pairs or groups to compute travel time and then individually to compute their age on arrival.
3. The formulas should be discussed before students do the activity page. The formula *rate x time = distance* is an important one that students can use

in a variety of situations. We use it as a division formula solving for time.

4. Teaching the process of crossing through an equal number of zeroes before dividing is very helpful. If students are dividing without calculators, this means they only need to divide by four.

5. When doing *Special Delivery,* students have to be reminded that a visit to any planet is not complete until they return to Earth.

Procedure

1. Ask the *Key Question* and state the *Learning Goals.*
2. Distribute the first student page and go over the information as a class.
3. Have students get into pairs or groups and distribute the second student page.
4. Tell them to record their ages in years and months in the appropriate space on the student page. They should round to the nearest month.
5. Give groups time to complete the calculations and fill in the table with the hours, days, months, and years it would take to get to the moon and each planet.
6. Have students individually calculate their ages at arrival for the moon and each planet. Discuss the process and students' results
7. Distribute the final student page an allow students time to complete the additional challenges and create some of their own.

Connecting Learning

1. How old would you be when you got to the moon? ...Mercury? ...Venus? ...Neptune?
2. Do these numbers surprise you? Why or why not?
3. What do these numbers tell you about our solar system?
4. How were you able to solve the problems on the special delivery page?
5. What mystery trip did you come up with?
6. Were other students able to solve your mystery trip? Why or why not?
7. What are you wondering now?

* Reprinted with permission from *Principles and Standards for School Mathematics,* 2000 by the National Council of Teachers of Mathematics. All rights reserved.

0006

OUT OF THIS WORLD

001001010110100100101010

EXTRA-TERRESTRIAL EXCURSIONS

001001010110100100101
04-171-979

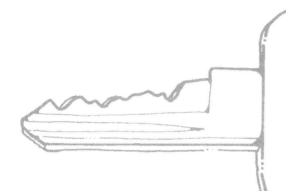

Key Question

How old would you be when you got to each planet in our solar system if you left Earth today, traveling at a speed of 40,000 kilometers per hour?

Learning Goals

Students will:

- determine the amount of time it would take to get to the moon and each planet in our solar system at a speed of 40,000 kilometers per hour, and
- use this information to determine how old they would be if they were able to travel to the planets at that speed.

The great distances in space are sometimes difficult to comprehend. If we look at the time it would take to travel to the moon and the planets by walking, by car, or by jet plane, we can begin to understand what a great undertaking interplanetary travel would be.

Time in Years from Earth

PLANETARY BODY	Walking—2.5 mph 3.6 km/h	Car—55 mph 80 km/h	Jet—990 mph 1436 km/h
Moon	11	.6	.03
Mercury	2588	133	7
Venus	1175	61	3
Mars	2222	113	6
Jupiter	17,843	909	46
Saturn	36,421	1848	92
Uranus	76,894	3935	194
Neptune	123,579	6289	313

EXTRA-TERRESTRIAL EXCURSIONS

0006

00100101011010010010101010

OUT OF THIS WORLD

00100101011010010010101010

04-171-979

Speed Limit: 40,000 km/h

Age today: _____ _____
            Years   Months

Average Distance From Earth / PLANETARY BODY ▼	HOURS — Distance / 40,000 (to nearest hour)	DAYS — Hours / 24 (to nearest hour)	MONTHS — Days / 30 (to nearest hour)	YEARS — Month / 12		ARRIVAL AGE — Years + Your Age	
				Years	Months	Years	Months
Moon 384,000 km							
Mercury 92,000,000 km							
Venus 41,000,000 km							
Mars 78,000,000 km							
Jupiter 629,000,000 km							
Saturn 1,227,000,000 km							
Uranus 2,721,000,000 km							
Neptune 4,347,000,000 km							

Imagine that you work for the Solar Systems Delivery Service. You need to determine the time necessary to make certain deliveries and return to Earth. The planets are not lined up in a straight line in their orbits around the sun. You must always return to Earth for refueling between planets.

1. Deliver communication systems to Mercury and Jupiter.

   Travel Time:_____

2. Deliver pizza to Venus and Mars.

   Travel Time:_____

3. You travel to one outer and one inner planet and back home again. Your journey takes you about 7 years, 6 months. To which planets did you travel?

   Planets:_____

4. Starting at Neptune, you travel home to Earth and then deliver letters to Mars.

   Travel Time:_____

5. Design a "Mystery" trip to two of the planets. Remember to always stop at the Earth between planets. How long would your mystery trip take?

   Travel Time:_____

REMEMBER, YOUR JOURNEY IS NOT OVER UNTIL YOU RETURN TO EARTH!

PLANETARY PIZZA

Exchange "Mystery" trips with another student and solve.

Connecting Learning

1. How old would you be when you got to the moon? ...Mercury? ...Venus? ...Neptune?

2. Do these numbers surprise you? Why or why not?

3. What do these numbers tell you about our solar system?

4. How were you able to solve the problems on the special delivery page?

5. What mystery trip did you come up with?

6. Were other students able to solve your mystery trip? Why or why not?

7. What are you wondering now?

Topic
Speed of radio waves

Key Question
How long would it take for someone on the moon or one of the planets to hear you say "hello"?

Learning Goal
Students will compute the time it would take for a message to reach the moon and the planets.

Guiding Documents
Project 2061 Benchmarks
- *Numbers and shapes—and operations on them—help to describe and predict things about the world around us.*
- *The earth is one of several planets that orbit the sun, and the moon orbits around the earth.*

NRC Standard
- *The earth is the third planet from the sun in a system that includes the moon, the sun, eight other planets and their moons, and smaller objects, such as asteroids and comets. The sun, an average star, is the central and largest body in the solar system.*

NCTM Standards 2000*
- *Understand the effects of multiplying and dividing whole numbers*
- *Develop and use strategies to estimate the results of whole-number computations and to judge the reasonableness of such results*
- *Solve problems that arise in mathematics and in other contexts*

Math
Computation
Estimation
   rounding
Problem solving
Applying formulas
Using calculators

Science
Earth science
   astronomy
      planets
Physical science
   radio waves

Integrated Processes
Observing
Comparing and contrasting
Recording data
Interpreting data
Predicting
Applying

Materials
Calculators
Student pages

Background Information
The distance to the moon and each planet in kilometers is listed on the student page. (Optional: These distances can be researched by the students prior to doing the activity.) Because radio waves travel at the speed of light, that is the rate used for transmission. The actual speed of light is 299,739 kilometers per second. It is rounded to 300,000 for ease in calculation since the answers should be rounded. Rounding strategies should be taught prior to this activity. Numbers may be rounded to the nearest tenth or whole number. The last computation, converting minutes to hours, should be done without a calculator, as the remainder will tell students the number of minutes.

Management
1. Time for this activity will vary depending on the students' math abilities and whether or not calculators are used.
2. Students can work in pairs or groups to do the computations.
3. The formula should be discussed before students do the activity pages. The formula, *rate x time = distance,* is an important one that students can use in a variety of situations. In this activity, it is used as a division formula solving for time.
4. Teaching the process of crossing out an equal number of zeroes before dividing makes the process much easier, as students then need only to divide by three. By doing this, students are actually first dividing both numbers by 100,000.

Procedure
1. Ask the *Key Question* and state the *Learning Goal.*
2. Discuss with the students how sound waves do not travel in space, so radio waves are used to transmit messages.

3. Have students read the information on the student page. Talk with them about rounding the speed at which radio waves travel (299,739 kilometers per second to 300,000 kilometers per second). Point out that the distances to the moon and the planets have also been rounded.

4. Invite students to use the formula on the student page to compute the time for a radio message to reach the moon and each planet.
   - Compute the number of seconds by dividing the distance from Earth by the speed of light. Round to the nearest whole number.
   - Compute the number of minutes by dividing the seconds by 60; then round to the nearest whole number.
   - Compute time in hours and minutes by dividing minutes by 60. This time, the remainder is used to determine the minutes.

Connecting Learning

1. Why did we round the numbers we were working with? [to make the calculations easier]

2. Does rounding give us exact answers to the question "How long would it take for someone on the moon or one of the planets to hear you say 'hello'?" Explain.

3. What do you notice about the differences in the amount of time it takes to communicate with the inner and outer planets?

4. What does that tell you about the distances to those planets?

5. What are you wondering now?

* Reprinted with permission from *Principles and Standards for School Mathematics*, 2000 by the National Council of Teachers of Mathematics. All rights reserved.

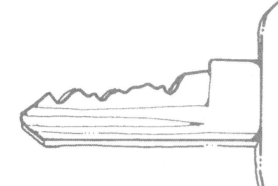

Key Question

How long would it take for someone on the moon or one of the planets to hear you say "hello"?

Learning Goal

Students will:

compute the time it would take for a message to reach the moon and the planets.

HOW LONG DOES IT TAKE TO SAY "HELLO"?

Although travel to the far reaches of space is a thing of the future, space communication is not. Exploatory space probes, such as Voyager 2, send and receive messages from the very edges of our solar system. Radio waves travel at the speed of light, 299,739 kilometers per second. Use the fomulas below to compute how long it would take for your "Hello" to reach the moon and each of the planets.

DESTINATION OF MESSAGE ▼	Average Distance from Earth in km	SECONDS Distance Rate (÷300,000)	MINUTES Seconds 60 (÷60)	HOURS Seconds 60 (÷60)
Moon	384,000			
Mercury	92,000,000			
Venus	41,000,000			
Mars	78,000,000			
Jupiter	629,000,000			
Saturn	1,227,000,000			
Uranus	2,721,000,000			
Neptune	4,347,000,000			

0007

OUT OF THIS WORLD

0010010101101001001010101010
HOW LONG DOES IT TAKE TO SAY "HELLO"?
0010010101101001001010
121-7-1985

Connecting Learning

1. Why did we round the numbers we were working with?

2. Does rounding give us exact answers to the question "How long would it take for someone on the moon or one of the planets to hear you say 'hello'?" Explain.

3. What do you notice about the differences in the amount of time it takes to communicate with the inner and outer planets?

4. What does that tell you about the distances to those planets?

5. What are you wondering now?

SPACE TALK MESSAGE

0008 001001010110100100101010   OUT OF THIS WORLD

001001010110100100101   0-3081-984

Topic
Computation

Key Question
How long would it take to send a series of communications to the moon or one of the planets?

Learning Goal
Students will compute the time it would take for a series of messages to go back and forth between the moon and the planets.

Guiding Documents
Project 2061 Benchmarks
- *Add, subtract, multiply, and divide whole numbers mentally, on paper, and with a calculator.*
- *Nine planets of very different size, composition, and surface features move around the sun in nearly circular orbits. Some planets have a great variety of moons and even flat rings of rock and ice particles orbiting around them. Some of these planets and moons show evidence of geological activity. The earth is orbited by one moon, many artificial satellites, and debris.*
- *The earth is one of several planets that orbit the sun, and the moon orbits around the earth.*

NRC Standards
- *The earth is the third planet from the sun in a system that includes the moon, the sun, eight other planets and their moons, and smaller objects, such as asteroids and comets. The sun, an average star, is the central and largest body in the solar system.*
- *Use mathematics in all aspects of scientific inquiry.*

NCTM Standards 2000*
- *Select appropriate methods and tools for computing with whole numbers from among mental computation, estimation, calculators, and paper and pencil according to the context and nature of the computation and use the selected method or tools*
- *Carry out simple unit conversions, such as from centimeters to meters, within a system of measurement*
- *Solve problems that arise in mathematics and in other contexts*

Math
Computation
Problem solving

Science
Earth science
  astronomy
    distances to planets

Integrated Processes
Observing
Collecting and recording data
Interpreting data
Applying

Materials
For the class:
  calculators
  student pages

Background Information
Students will have computed the time for a one-way transmission in the activity *How Long Does it Take to Say Hello?* Now they will use those calculations to compute the time necessary to complete a series of two-way communications. You may use the six-sentence script that is provided, or have students create their own scripts of varying lengths. It works best if students multiply the time for a one-way transmission in minutes (seconds for the moon) by the number of transmissions made. They will then divide any answers over 60 minutes in length by 60 to compute hours and minutes to determine at what time and day they will complete their transmissions.

Management
1. Time for this activity will vary depending on the students' math abilities and whether or not calculators are used. It will also depend on whether students write their own transmission scripts and take the time to actually try them.
2. Students compute by adding the time of the total transmission to the day and time they began.
3. Students can work in pairs or collaborative groups.
4. Students need to use their answers for one-way transmissions from *How Long Does it Take to Say Hello?* They then answer the question, "If you began transmitting at 8:00 A.M. Tuesday, what day and time would you finish?"

Procedure

1. Ask the *Key Question* and state the *Learning Goals*.
2. Have students get into pairs or groups and distribute the student pages.
3. Give groups time to complete the calculations and fill in the table with the total transmission time to the moon and each planet. If that answer is greater than 60 minutes, it should be divided by 60 to give the answer in hours and minutes. This will be true for the planets Jupiter through Neptune.
4. Close with a time of class discussion.

Connecting Learning

1. How long would it take to complete the transmission to the moon? ...Mercury? ...Venus? ...Neptune?
2. Do these numbers surprise you? Why or why not?
3. What do these numbers tell you about our solar system?
4. Between which two planets did you see a change from time in minutes to hours?
5. Scientists sometimes refer to inner and outer planets. From what you have learned so far, which planets do you think are inner planets? [Mercury, Venus, Earth, Mars] ...outer planets? [Jupiter, Saturn, Uranus, Neptune] Explain your reasoning.
6. What are you wondering now?

* Reprinted with permission from *Principles and Standards for School Mathematics*, 2000 by the National Council of Teachers of Mathematics. All rights reserved.

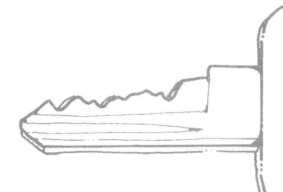

Key Question

How long would it take to send a series of communications to the moon or one of the planets?

Learning Goal

Students will:

compute the time it would take for a series of messages to go back and forth between the moon and the planets.

© 2007 AIMS Education Foundation

You need to send the following important message to Space Bases on the moon and the planets and to receive their responses. Use the transmission time from "How Long Does it Take to Say 'Hello'?" to determine how long each set of transmissions will take.

Divide yourselves into small groups and time yourselves to see how much of this transmission can be sent and received in your school day. Use the *Space Talk* activity page "Time to Transmit" to see how long each set would take to complete.

Write your own set of messages sent and received. Determine how long it would take to complete them.

SPACE TALK

OUT OF THIS WORLD

TIME TO TRANSMIT...

0-3081-984

Use the data from "How Long Does it Take to Say 'Hello'?" to compute the total transmission time for your message.

IF YOU BEGAN TRANSMITTING ON TUESDAY AT 8:00 A.M., WHAT DAY AND TIME WOULD YOU FINISH?

Time at Completed Transmission

DAY	TIME

PLANET ▶	One-Way Transmission Time from Earth	Number of Transmissions	Total Transmission Time	
Moon		×	=	minute: / seconds:
Mercury		×	=	hours: / minutes:
Venus		×	=	hours: / minutes:
Mars		×	=	hours: / minutes:
Jupiter		×	=	hours: / minutes:
Saturn		×	=	hours: / minutes:
Uranus		×	=	hours: / minutes:
Neptune		×	=	hours: / minutes:

Connecting Learning

1. How long would it take to complete the transmission to the Moon? ...Mercury? ...Venus? ...Neptune?

2. Do these numbers surprise you? Why or why not?

3. What do these numbers tell you about our solar system?

4. Between which two planets did you see a change from time in minutes to hours?

5. Scientists sometimes refer to inner and outer planets. From what you have learned so far, which planets do you think are inner planets? ...outer planets? Explain your reasoning.

6. What are you wondering now?

Topic
Computation

Key Question
What do you think you would weigh on the moon? ...on each planet in our solar system?

Learning Goal
Students will determine what their weight would be on the moon and each of the planets in our solar system.

Guiding Documents
Project 2061 Benchmarks
- *Things on or near the earth are pulled toward it by the earth's gravity.*
- *Tables and graphs can show how values of one quantity are related to values of another.*
- *Add, subtract, multiply, and divide whole numbers mentally, on paper, and with a calculator.*
- *Mathematical ideas can be represented concretely, graphically, and symbolically.*

NRC Standards
- *Gravity is the force that keeps planets in orbit around the sun and governs the rest of the motion in the solar system. Gravity alone holds us to the earth's surface and explains the phenomena of the tides.*
- *Use mathematics in all aspects of scientific inquiry.*

NCTM Standards 2000*
- *Select appropriate methods and tools for computing with whole numbers from among mental computation, estimation, calculators, and paper and pencil according to the context and nature of the computation and use the selected method or tools*
- *Represent and analyze patterns and functions, using words, tables, and graphs*
- *Represent data using tables and graphs such as line plots, bar graphs, and line graphs*

Math
Computation
Data analysis
  bar graph
  tables

Science
Earth science
  astronomy
    planets
Physical science
  surface gravity

Integrated Processes
Observing
Predicting
Comparing and contrasting
Collecting and recording data
Interpreting data
Applying

Materials
For the class:
  calculators
  student pages

Background Information
The gravity of an object is related to its mass and density. The more mass an object has, the stronger its pull of gravity. Gravitational pull weakens with distance. The mean densities of the outer planets are less than Earth. (their interior densities, however, are far greater.) Saturn actually has a mean density less than water.

The surface gravity of Earth is considered as one, and the moon and the other planets are considered as a fraction of that. Each student will multiply the estimated surface gravity for the specific planet times his or her weight in pounds (or kilograms) to determine their weight on the various planets. Information about the surface gravity of some planets varies. Students should be reminded that our information about space is constantly being updated.

Management
1. This activity will take about 30-40 minutes.
2. Students can either be weighed in class or bring their recorded weights from home.
3. If students are sensitive about their weights, use the weight of an object in the classroom.

Procedure

1. Ask the *Key Question* and state the *Learning Goals.*
2. Distribute the student pages and calculators. Go over the information about gravity on the first student page.
3. Have students record their weights on Earth and their predicted weights on the moon and each planet in the table.
4. Give students time to complete the second student page using the information from the table on the first page. Once they have determined their weights on each planetary body, have them complete the table on the first page and answer the questions.
5. Have students use the information from the tables to make a bar graph on the third student page.
6. Close with a time of class discussion.

Connecting Learning

1. On which planet would you weigh the least? ...the most?
2. On which planets would you weigh more than on Earth? [Jupiter, Saturn, Neptune] How do you know? [their surface gravities are greater than Earth's]
3. On which planets would you weigh nearly the same as you do on Earth? [Venus, Saturn, Uranus] How do you know? [their surface gravities are close to 1]
4. For which planet was your prediction the closest? Why do you think this is?
5. For which planet was your prediction the furthest from the actual weight? Why do you think this is?
6. What does the graph tell you about the data?
7. What are you wondering now?

*   Reprinted with permission from *Principles and Standards for School Mathematics*, 2000 by the National Council of Teachers of Mathematics. All rights reserved.

> ● **Key Question** ●
> What do you think you
> would weigh on the moon?
> ...on each planet in our
> solar system?

Learning Goal

Students will:

determine what their weight would be on
the moon and each of the
planets in our solar system.

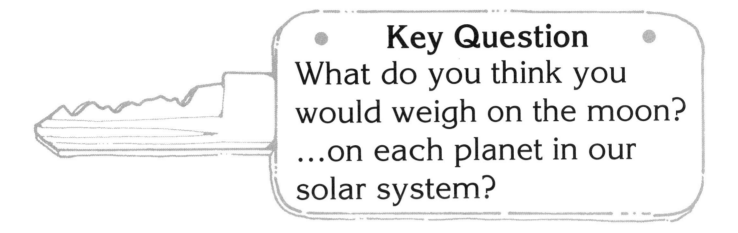

WEIGHT IN SPACE

Gravity is an invisible force that pulls on things. The pull of Earth's gravity is what gives us weight. The size, mass, and density of a planet or moon determines its gravitational pull. We consider the surface gravity of the Earth to be 1. Look at the chart of surface gravities of the moon and other planets and predict what you think your weight would be on each of them.

My weight on Earth is:_____.

PLANETARY BODY ▼	Estimated Surface Gravity	Predicted Weight	Actual Weight	Difference
Moon	.16			
Mercury	.39			
Venus	.91			
Mars	.38			
Jupiter	2.53			
Saturn	1.07			
Uranus	.91			
Neptune	1.15			

On which planet would you weigh the least?_____

On which planets would you weigh more than on the Earth?_____

On which planets would you weigh nearly the same as on Earth?

_____

To find your actual weight on the moon and the planets, you must multiply your weight on Earth by the surface gravity of that body. Find your weight on the moon first and then follow the formula to complete the chart below.

Use the surface gravities from the chart on the previous page. Round your answer to the nearest whole number.

The surface gravity on the moon is one-sixth or .16 that of Earth.

PLANETARY BODY	My Weight on Earth X	Surface Gravity	=	My Weight on Planetary Body
Moon	X		=	
Mercury	X		=	
Venus	X		=	
Mars	X		=	
Jupiter	X		=	
Saturn	X		=	
Uranus	X		=	
Neptune	X		=	

For which planet was your prediction the closest?_____

For which planet was your prediction the most different?_____

On which planet was your weight closest to that on Earth?_____

WEIGHT IN SPACE

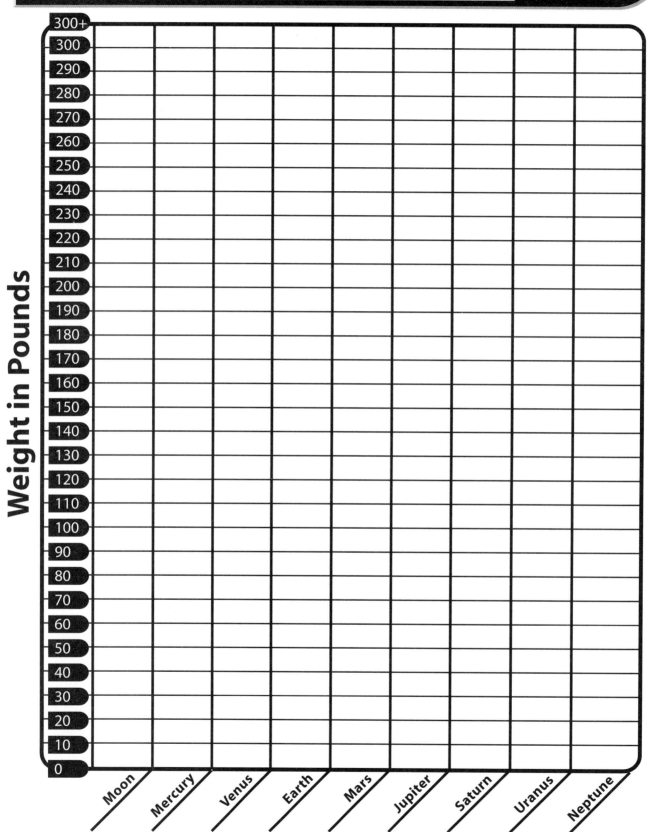

Weight in Pounds

	Moon	Mercury	Venus	Earth	Mars	Jupiter	Saturn	Uranus	Neptune
300+									
300									
290									
280									
270									
260									
250									
240									
230									
220									
210									
200									
190									
180									
170									
160									
150									
140									
130									
120									
110									
100									
90									
80									
70									
60									
50									
40									
30									
20									
10									
0									

138

WEIGHT IN SPACE

Connecting Learning

1. On which planet would you weigh the least? ...the most?

2. On which planets would you weigh more than on Earth? How do you know?

3. On which planets would you weigh nearly the same as you do on Earth? How do you know?

4. For which planet was your prediction the closest? Why do you think this is?

5. For which planet was your prediction the furthest from the actual weight? Why do you think this is?

6. What does the graph tell you about the data?

7. What are you wondering now?

Topic
Computation

Key Question
How far or high can you jump and throw on the moon and the planets in our solar system?

Learning Goal
Students will compute gravity factors to determine how far and high they could jump and throw on a planetary body.

Guiding Documents
Project 2061 Benchmarks
- *Things on or near the earth are pulled toward it by the earth's gravity.*
- *Add, subtract, multiply, and divide whole numbers mentally, on paper, and with a calculator.*

NRC Standards
- *Gravity is the force that keeps planets in orbit around the sun and governs the rest of the motion in the solar system. Gravity alone holds us to the earth's surface and explains the phenomena of the tides.*
- *Use mathematics in all aspects of scientific inquiry.*

NCTM Standard 2000*
- *Select appropriate methods and tools for computing with whole numbers from among mental computation, estimation, calculators, and paper and pencil according to the context and nature of the computation and use the selected method or tools*

Math
Multiplication
Data analysis
  averages
  rounding
Measurement
  length
  time
Problem solving

Science
Earth science
  astronomy
    planets
Physical science
  surface gravity

Integrated Processes
Observing
Comparing and contrasting
Collecting and recording data
Interpreting data
Applying

Materials
For the class:
  butcher paper
  masking tape
  inkpad
  paper towels
  measuring tape
  stopwatch
  sponge-like ball
  calculators
  student pages

Background Information
Gravity is an invisible force that pulls on objects. All the planets in our solar system have a gravitational pull. The gravitational pull of each planet is different because of its mass and density. The surface gravity of Earth is considered to be *one*. Students will use the surface gravity of the moon and planets to compute *gravity factors* by multiplying the gravity factor by any measurable quantity from an activity performed on Earth, they can determine their performance on other planetary bodies.

Management
1. Students work in small groups.
2. Set up four centers around the room: *Quantum Leap, Gravity Defiance, Planetary Pitch,* and *Asteroid Throw.*
3. At the *Quantum Leap* center, mark off a starting place with masking tape so that students can make a standing long jump. Place two or three measuring tapes at this location.
4. At the *Gravity Defiance* center, tape a sheet of butcher paper to the wall at a height you estimate students can reach when they jump up. Leave one tape measure, the ink pad, and paper towels for ink cleanup at this center.
5. At the *Planetary Pitch* center, mark off a starting line with masking tape in an area where students can throw a ball. Leave several measuring tapes and a sponge-like ball.

6. For *Asteroid Throw,* set up a center outside where students can throw a ball up into the air. Leave a stopwatch and ball at this center.

7. Prior to the activity, post the rules for the *Galactic Games* at the various locations where the events will be held.

Procedure

1. Ask the *Key Question* and state the *Learning Goals.*
2. Tell students that they are going to record their performance in several activities to see how they would perform on different planets.
3. Distribute the first student page and have students determine the gravity factor for the moon and each of the planets. Have them identify the sets of planets that have very similar gravity factors.
4. Describe the four centers and the task students will be performing at each one (see *Rules).* Distribute the second student page to each student and go over what they need to record.
5. Have students get into their groups and assign each group to a center. Allow them time to complete each task and rotate through all four centers.
6. When all groups are finished, distribute the third student page and have them complete the table with their performance on the moon and each planet.
7. Close with a time of class discussion. Compare the results from the different groups.

Connecting Learning

1. On which planetary body would you do the best at the *Planetary Pitch*? ...the worst? Why?
2. On which planetary bodies would your performance be closest to your performance here on Earth? Why?
3. What would be the difference between your best and worst performance in the *Asteroid Throw*?
4. If you were on Venus and your teammates were on Earth, who would win the *Quantum Leap*? Why? What if you were on Jupiter?
5. What are you wondering now?

* Reprinted with permission from *Principles and Standards for School Mathematics,* 2000 by the National Council of Teachers of Mathematics. All rights reserved.

Key Question

How far or high can you jump and throw on the moon and the planets in our solar system?

Learning Goal

Students will:

compute gravity factors to determine how far and high they could jump and throw on a planetary body.

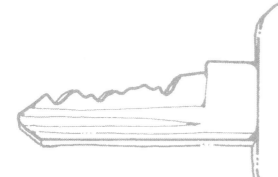

142

Rules for Quantum Leap

1. Put both of your feet on the starting line. Jump as far as you can.

2. Measure the distance from the starting line to the place you landed. You must measure to the closest place touched by any part of your body.

3. Record this distance and repeat two more times.

Rules for Gravity Defiance

1. Mark your index finger using the inkpad.

2. Reach as high as you can and mark the paper. This is your *base mark*.

3. Again, mark your index finger using the inkpad.

4. Jump up and touch the paper as high as you can. This is your *jump mark*.

5. Measure the distance between your *base mark* and your *jump mark*.

6. Record this distance and repeat two more times.

Rules for Planetary Pitch

1. With both feet on or behind the starting line, throw the ball as far as you can.

2. Measure the distance from where the ball first hits the ground to where you were standing on the starting line.

3. Record this distance and repeat two more times.

Rules for Asteroid Throw

Timer:

1. Tell the competitor to "go" and start the stopwatch as soon as the ball leaves his or her hand.

2. Stop the stopwatch as soon as the ball hits the ground.

3. Tell the competitor the time.

Competitor:

1. When the timer says "go," throw the ball in the air as high as you can.

2. Record the time it stays in the air and repeat two more times.

Gravity is an invisible force that pulls on objects. Each of the planets in our solar system has a gravitational pull that is different because of its mass and density. The surface gravity of Earth is considered to be 1. Each planet has a different surface gravity. Athletic events on other planets would have much different outcomes. To find the gravity factor for each of the planets, divide the surface gravity of Earth (1) by the surface gravity of each planet. Find the gravity factor for each planet.

Use this gravity factor to determine how long and how high you can jump and throw on the other planets.

PLANETARY BODY ▼	Gravity on Earth	÷	Surface Gravity	=	Gravity Factor
Moon	1	÷	.16	=	
Mercury	1	÷	.39	=	
Venus	1	÷	.91	=	
Mars	1	÷	.38	=	
Jupiter	1	÷	2.53	=	
Saturn	1	÷	1.07	=	
Uranus	1	÷	.91	=	
Neptune	1	÷	1.15	=	

ROUND GRAVITY TO THE NEAREST TENTH!

On which pairs of planets is the gravity factor almost the same?

1. _____   _____

2. _____   _____

3. _____   _____

Compete in each event.
Make three trials and find the average.

GALACTIC GAMES

★1 QUANTUM LEAP

TRIALS
1.
2.
3.
TOTAL
AVERAGE

★2 GRAVITY DEFIANCE

TRIALS
1.
2.
3.
TOTAL
AVERAGE

★3 PLANETARY PITCH

TRIALS
1.
2.
3.
TOTAL
AVERAGE

★4 ASTEROID THROW

TRIALS
1.
2.
3.
TOTAL
AVERAGE

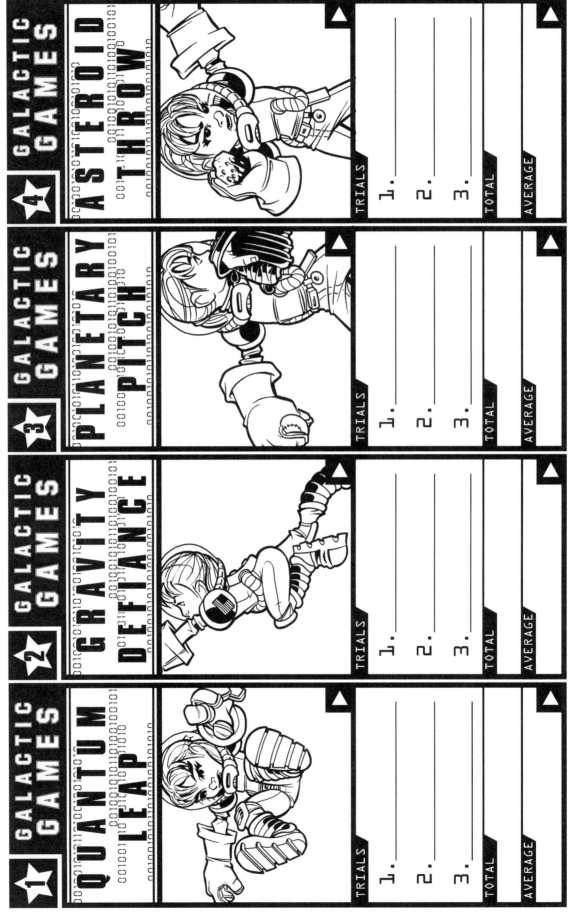

GALACTIC GAMES

OUT OF THIS WORLD

0-1201-977

0010

FORMULA: Average X Gravity Factor = Distance

GALACTIC GAME ▼	Earth Average	Moon x 6.3	Mercury & Mars x 2.6	Venus & Uranus x 1.1	Jupiter x .4	Saturn & Neptune x .9
QUANTUM LEAP						
GRAVITY DEFIANCE						
PLANETARY PITCH						
ASTEROID THROW						

0010

OUT OF THIS WORLD

GALACTIC GAMES

0010010101101001001010

0010010101101001001010

0-1201-977

Connecting Learning

1. On which planetary body would you do the best at the *Planetary Pitch*? ...the worst? Why?

2. On which planetary bodies would your performance be closest to your performance here on Earth? Why?

3. What would be the difference between your best and worst performanace in the *Asteroid Throw*?

4. If you were on Venus and your teammates were on Earth, who would win the *Quantum Leap*? Why? What if you were on Jupiter?

5. What are you wondering now?

0011   0010010101101001001010

PLANET TRIVIA

OUT OF THIS WORLD

0010010101101001001010 0308-198-1

Topic
Planets

Key Question
How can you use what you have learned about the planets to win the *Planet Trivia* game?

Learning Goal
Students will play *Planet Trivia* to reinforce and review what they have learned about the planets.

Guiding Documents
Project 2061 Benchmarks
- *Nine planets of very different size, composition, and surface features move around the sun in nearly circular orbits. Some planets have a great variety of moons and even flat rings of rock and ice particles orbiting around them. Some of these planets and moons show evidence of geological activity. The earth is orbited by one moon, many artificial satellites, and debris.*
- *The earth is one of several planets that orbit the sun, and the moon orbits around the earth.*
- *We live on a relatively small planet, the third from the sun in the only system of planets definitely known to exist (although other, similar systems may be discovered in the universe).*

NRC Standard
- *The earth is the third planet from the sun in a system that includes the moon, the sun, eight other planets and their moons, and smaller objects, such as asteroids and comets. The sun, an average star, is the central and largest body in the solar system.*

Science
Earth science
  astronomy
    planets

Integrated Processes
Observing
Inferring
Applying

Materials
For each group:
  one set of trivia cards
  directions

For each student:
  one blank page to write additional questions

Background Information
Most of the answers can be found in the information provided in this book. A few questions such as, "What star is closest to the Earth?" [the sun], are general knowledge questions. Students should be encouraged to write questions of their own that require extra research.

Management
1. Play time will vary depending on whether play is done individually or by teams and the number of players there are.
2. Divide students into groups. Students can play as individuals in groups of two to eight students or as teams of two to four players each.
3. Each group needs a set of cards and a rule sheet.
4. The player or team with the most cards at the end of the playing time wins.

Procedure
1. Have students get into groups and distribute one set of game cards and a rule sheet to each group.
2. Go over the instructions for the game as a class and answer any questions.
3. Provide time for groups to play several rounds of the game.
4. Allow students to write additional questions on the blank cards provided and add those to the game.

Connecting Learning
1. Which questions were the most difficult to answer? Why?
2. Which questions were the easiest to answer? Why?
3. What questions did you add to the game? How did you come up with these questions?
4. What are you wondering now?

Key Question

How can you use what you have learned about the planets to win the *Planet Trivia* game?

Learning Goal

Students will:

play *Planet Trivia* to reinforce and review what they have learned about the planets.

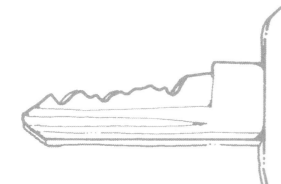

Two to eight players for each set of cards.

INDIVIDUAL SCORES

Deal six cards face down to each player. This is your hand for the first round of *Planet Trivia*. Set any extra cards aside.

The person to the right of the dealer begins by asking the person on his or her right a question. If the correct answer is given, the player giving the answer takes the card, keeping it separate from his or her hand. If an incorrect answer is given, the person asking the question reads the correct answer and places the card in the discard pile at the center of the table. Play passes to the right until everyone has used their six cards. The player with the most cards at the end of the game is the winner. To play again, shuffle and redeal.

TEAM SCORES

Shuffle the cards and deal each team 16 cards face down. Set remaining cards aside. Teams take turns asking questions of the opposing team. Teams are allowed 20 seconds to confer before giving an answer, but only one answer may be given. When a team answers correctly, it keeps the card. If an incorrect answer is given, the correct answer is read and the question is placed in the discard pile. Play continues until both teams use all 16 cards. The team with the most cards wins.

PLANET TRIVIA CARDS

Q: Which planet is farthest from the sun?

A: Neptune

Q: Which planet has the most moons?

A: Saturn

Q: Which two planets have no moons?

A: Venus and Mercury

Q: Which planet has the shortest year?

A: Mercury

Q: Which planet has the longest period of rotation?

A: Venus

Q: What star is closest to Earth?

A: The sun

Q: Which planet travels around the sun once every 165 Earth years?

A: Neptune

Q: What is the largest planet in the solar system?

A: Jupiter

Q: Which planet is considered to be Earth's twin in mass and size?

A: Venus

Q: What is the second largest planet in the solar system?

A: Saturn

Q: What is the astronomical term for the distance light travels in a year?

A: A light year

Q: What fraction of your weight would you weigh on the moon?

A: One sixth (1/6)

Q: What planet became a "dwarf planet" in 2006?

A: Pluto

Q: Which planet is closest to the sun?

A: Mercury

Q: Which planet has the largest mountain in the solar system?

A: Mars

Q: What mythological person was the planet closest to the Earth named for?

A: Venus—goddess of love

PLANET TRIVIA CARDS 2

Q: What is the largest satellite orbiting the Earth?

A: The moon

Q: What are the solar system's planets named for?

A: Ancient gods

Q: Which planet was discovered beyond the orbit of Saturn in 1781?

A: Uranus

Q: What is Earth's home galaxy called?

A: The Milky Way

Q: What space probe photographed the rings of Uranus?

A: Voyager II

Q: What planet is often referred to as the morning and evening star?

A: Venus

Q: Which planet is called the Red Planet?

A: Mars

Q: Which planet is brightest as seen from Earth?

A: Venus

Q: Which planet boasts the Great Red Spot?

A: Jupiter

Q: Which planets did the Mariner spacecraft explore?

A: Mercury and Mars

Q: Which planet's day is closest in length to Earth's?

A: Mars

Q: Which planet has a day that is longer than its period of revolution?

A: Mercury

Q: Which planet has a surface gravity closest to that on Earth?

A: Saturn

Q: Which planet has the greatest mass?

A: Jupiter

Q: On which planet would you weigh the least?

A: Mars

Q: Which planet has a density less than water?

A: Saturn

PLANET TRIVIA CARDS 3

Q: Which planets are larger than Earth?

A: Jupiter, Saturn, Uranus, Neptune

Q: Which planets are smaller than Earth?

A: Mercury, Mars, Venus

Q: What are the names of the outer planets?

A: J

Q: What are the names of the inner planets?

A: Mercury, Venus, Earth, Mars

Q: Which planets are the Gas Giants?

A: Jupiter, Saturn, Uranus, Neptune

Q: How many months does it take for the moon to revolve around the Earth?

A: less than one—about 28 days

Q: Which planet has moons named after Shakespeare characters?

A: Uranus

Q: How fast can we radio messages through space?

A: the speed of light

Q: What is an astronomical unit equivalent to?

A: Earth's distance from the sun

Q: What is the name of the unit astronomers use to measure nearby space distances?

A: Astronomical unit

Q: Which planet has the fastest winds in the solar system?

A: Neptune

Q: Which planet is sometimes called the water, or blue planet?

A: Earth

Q: What planet is named for the god of the sea and what is its symbol?

A: Neptune, Trident

Q: Which planet spins on its side?

A: Uranus

Q: Which planet has "ears"?

A: Saturn

Q: Which planet is twice as massive as all the other planets combined?

A: Jupiter

PLANET TRIVIA CARDS 4

PLANET TRIVIA

Connecting Learning

1. Which questions were the most difficult to answer? Why?

2. Which questions were the easiest to answer? Why?

3. What questions did you add to the game? How did you come up with these questions?

5. What are you wondering now?

Topics
Geometry, communication

Key Question
How can you follow instructions exactly to place shapes on a paper in a specific arrangement?

Learning Goals
Students will:
- experience how space voice communication worked early in the space program,
- familiarize themselves with geometric shapes,
- learn to give clear instructions, and
- practice listening carefully to their peers and following instructions precisely.

Guiding Documents
Project 2061 Benchmark
- *Communication technologies make it possible to send and receive information more and more reliably, quickly, and cheaply over long distances.*

NCTM Standards 2000*
- *Identify, compare, and analyze attributes of two- and three-dimensional shapes and develop vocabulary to describe the attributes*
- *Describe location and movement using common language and geometric vocabulary*
- *Communicate their mathematical thinking coherently and clearly to peers, teachers, and others*

Math
Geometry
Spatial sense

Integrated Processes
Observing
Communicating
Comparing and contrasting
Collecting and recording data
Interpreting data
Applying

Materials
For each student:
one set of geometric shapes
one 8.5" x 11" piece of paper
scissors

For the class:
transparency of the geometric shapes
box lids for dividers (see *Management 3*)

Management
1. This activity is divided into two parts. In the first part, students participate in a whole-class experience where they practice following instructions given by the teacher. In the second part, they create their own instructions and give them to a classmate to follow.
2. Copy a set of the geometric shapes on overhead transparency film and cut them out ahead of time.
3. If possible, use the lids from paper boxes to make dividers that can be placed between students' desks in *Part Two*. This will allow them to face each other, but not see each other's papers. Shoe boxes and lids, poster board, or cereal boxes would also work. If dividers are not available, students can sit back to back, but this makes giving and listening to instructions more difficult.

Procedure
Part One
1. Ask the *Key Question* and state the *Learning Goals*.
2. Distribute the shape pages, scissors, and a piece of paper to be used as the instrument panel to each student. Have them carefully cut out all of the shapes.
3. Explain to the students that when the space program first began, spacecraft did not have television cameras. Mission control and the astronauts communicated without seeing each other. They relied on verbal communication, so it was critical that the information they gave be detailed and correct.
4. Tell students that they will be simulating an experience that an early astronaut might have had when communicating with mission control. Explain that you will be giving them a series of detailed verbal instructions, and that you want them to follow those instructions exactly.
5. Read the following instructions to the class, giving them time to complete the instructions after each step. You may repeat the instructions, but do not allow any questions for clarification. As you read each instruction, put the correct shape in the specified location on the overhead projector, but do not turn the projector on.

- Place your instrument panel (piece of paper) in front of you so the longer side of the rectangle is a the bottom.
- Place the large square in the upper left-hand corner of the instrument panel.
- Place the large rectangle along the right side of the square so its longer side touches the top of the instrument panel.
- Place the large circle so it touches the large rectangle at the mid-point of the rectangle's base.
- Place the large hexagon in the lower right-hand corner of the instrument panel so one side rests on the bottom of the rectangle.
- Place the small ellipse in the center of this large hexagon.
- Place the small rectangle along the top of the hexagon so it touches the right side of the instrument panel.

7. After you have finished reading the instructions, distribute the first student page. Turn on the overhead projector so that students can see the correct arrangement of the shapes, and have them complete the first question.

Part Two

1. Have students get into pairs and set up a divider between each pair. Tell them that it is important to not look at each other's papers during this activity.
2. Instruct the pairs to decide who will be mission control and who will be the astronaut.
3. Have the mission control students arrange some of the shapes on their instrument panels. Tell them that they must use at least eight of the shapes, but can use more than eight if they wish to.
4. When they are ready, have the students who are mission control give instructions to the astronauts so that the astronauts can arrange their shapes in the same way. Remind students that they can only use verbal communication and cannot look at each other's papers. The astronauts cannot ask questions of mission control at any time.
5. Once students have completed their instructions, have students remove the divider and compare their papers. Have the player who was the astronaut record his or her results on the student page.
6. Tell the students to reverse roles and repeat the process, once again without asking questions.
7. Have pairs repeat the process again, this time, with the astronaut being allowed to ask clarifying questions of mission control.
8. Distribute the final student page and have students complete the questions. Close with a time of class discussion and sharing.

Connecting Learning

1. How accurate were you at placing your shapes during the whole-class trial? How did this compare to your classmates?
2. Was it easier or harder when you did the second trial (no questions) with your partner? Why?
3. Was it easier to be the astronaut or mission control? Why?
4. How did the third trial (questions allowed) compare to the first two?
6. Did you do better as the astronaut or as mission control? How do you know?
7. What would you do to make communication better between the astronaut and mission control?
8. What are you wondering now?

Extensions

1. Make multiple copies of the geometric shapes in different colors to add the variable of color into the problem.
2. Have students give verbal instructions for carrying out a specific task, such as folding a paper airplane, to gain additional practice in communicating clearly and effectively.

* Reprinted with permission from *Principles and Standards for School Mathematics*, 2000 by the National Council of Teachers of Mathematics. All rights reserved.

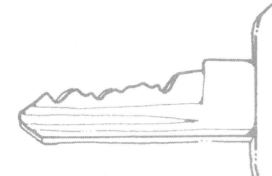

Key Question

How can you follow instructions exactly to place shapes on a paper in a specific arrangement?

Learning Goals

Students will:

- experience how space voice communication worked early in the space program,
- familiarize themselves with geometric shapes,
- learn to give clear instructions, and
- practice listening carefully to their peers and following instructions precisely.

0012

OUT OF THIS WORLD

PHONE HOME
GEOMETRIC SHAPES

0010010101101001001010

012-9194-9

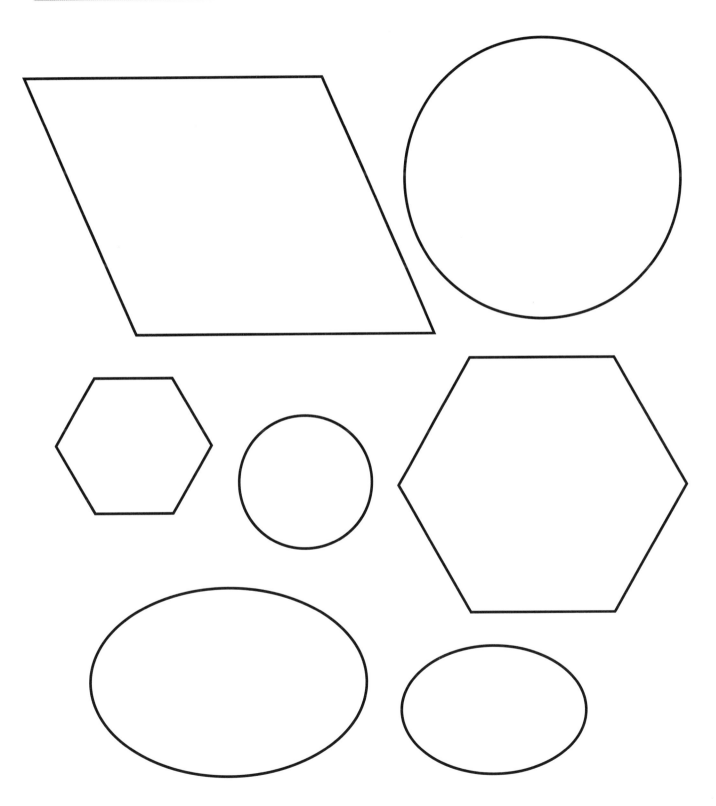

PHONE HOME

012-9194-9

1. You are an astronaut receiving instructions. Listen to the directions carefully. Try to place your shapes correctly without asking any questions.

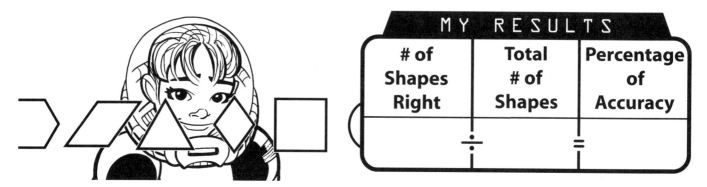

MY RESULTS		
# of Shapes Right	Total # of Shapes	Percentage of Accuracy
÷	=	

2. Next, work with a partner. Take turns being the astronaut and Mission Control. Listen to your partner's instruction and try to place your shapes correctly without asking questions. Record each person's results as the astronaut.

WITHOUT QUESTIONS

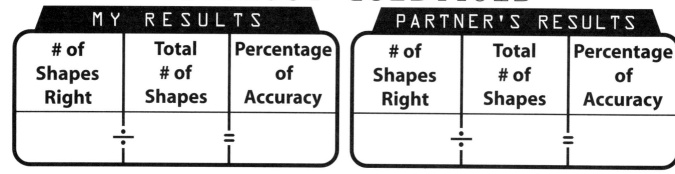

MY RESULTS		
# of Shapes Right	Total # of Shapes	Percentage of Accuracy
÷	=	

PARTNER'S RESULTS		
# of Shapes Right	Total # of Shapes	Percentage of Accuracy
÷	=	

3. Try it one more time. This time, you may ask questions when given directions. Record each astronaut's results.

WITH QUESTIONS

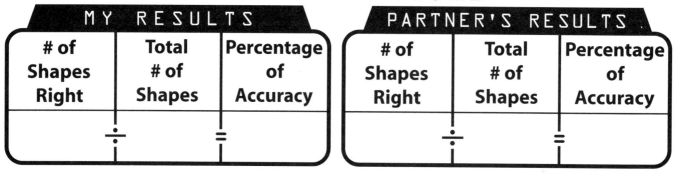

MY RESULTS		
# of Shapes Right	Total # of Shapes	Percentage of Accuracy
÷	=	

PARTNER'S RESULTS		
# of Shapes Right	Total # of Shapes	Percentage of Accuracy
÷	=	

1. Who had the highest percentage of accuracy?_____

2. How many figures did you place correctly?
   Trial 2:_____   Trial 3:_____

3. On a scale of 1 to 10, how hard was it to be Mission Control?

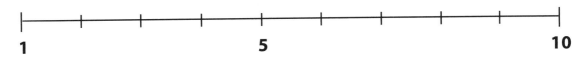

4. On a scale of 1 to 10, how hard was it to be the astronaut?

5. Which of the two was harder for you? Why?

6. On a scale of 1 to 10, how hard was Trial 2? (no questions)

7. On a scale of 1 to 10, how hard was Trial 3? (with questions)

8. Which did you like better, being Mission Control or being the astronaut? Why?

9. What would you do to make communication better between the astronaut and Mission Control?

PHONE HOME

Connecting Learning

1. How accurate were you at placing your shapes during the whole-class trial? How did this compare to your classmates?

2. Was it easier or harder when you did the second trial (no questions) with your partner? Why?

3. Was it easier to be the astronaut or mission control? Why?

4. How did the third trial (questions allowed) compare to the first two?

6. Did you do better as the astronaut or as mission control? How do you know?

7. What would you do to make communication better between the astronaut and mission control?

8. What are you wondering now?

Topic
Data collection

Key Question
What items would you send to another planet to exemplify how we live on planet Earth?

Learning Goal
Students will determine what items they would send in a time capsule to extraterrestrial life on another planet.

Guiding Document
NCTM Standards 2000*

- *Collect data using observations, surveys, and experiments*
- *Represent data using tables and graphs such as line plots, bar graphs, and line graphs*
- *Solve problems that arise in mathematics and in other contexts*

Math
Data collection
Graphing

Integrated Processes
Observing
Collecting and recording data
Interpreting data
Applying

Materials
Scissors
16 envelopes
Student pages

Background Information
In 1977, NASA launched the Voyager 1 and 2 spacecraft. Both are currently much farther from the sun than Pluto is, moving away from our solar system at more than 17 kilometers per second. Aboard each one of these spacecraft is a sort of time capsule intended to communicate information about Earth to any being that might someday come across it. The gold-plated copper phonograph records have sounds and images chosen to represent nature, people, and places on Earth. For more information on the golden records, as they are known, visit the Jet Propulsion Laboratory's Voyage website at http://voyager.jpl.nasa.gov/spacecraft/goldenrec.html

Management
1. This activity may take one to two days to complete.
2. Reproduce two copies of *Space Capsule* for each student. One is for individual results and the second is for small-group results.
3. Label the 16 envelopes with one each of the listed categories.
4. Three to four students will work together in groups.

Procedure
1. Introduce the activity by reading the following scenario (or creating a similar one). Let's imagine that we have just discovered that intelligent beings exist on another planet in our galaxy and we want to communicate with them. Because the distance between our planets is so great, it is impossible for us to travel there. Instead, we are going to send a rocket with a time capsule containing 16 well-chosen items that will exemplify life on planet Earth. What should the capsule contain?
2. Distribute one copy of the response page to each student and give them time to complete their selections for each category.
3. Have the students get into groups and discuss the responses they each recorded for the different categories. Give each group a second copy of the response page and have them agree on one response for each category.
4. Display the labeled envelopes at the front of the class.
5. Instruct students to write the number of people in their group on each category strip, cut apart the strips, and place the responses in the correct envelope.
6. Once all groups have put their responses in the envelopes, give a few envelopes to each group to sort. Have them identify the top three responses in each category and the number of students who selected that item.
7. Distribute the two copies of the final student page to each student and have groups share their information so that everyone can graph the results.

Connecting Learning

1. How did you decide what to put for each category?
2. How did your group come to an agreement on one response for each category?
3. Do you think this was a fair method to use? Why or why not?
4. Are you happy with the responses your group selected? Why or why not?
5. How did your group's responses compare to the other groups?
6. How do you think our classes' responses would compare to students in the same grade somewhere else in the country or the world?
7. What are you wondering now?

* Reprinted with permission from *Principles and Standards for School Mathematics*, 2000 by the National Council of Teachers of Mathematics. All rights reserved.

SPACE CAPSULE

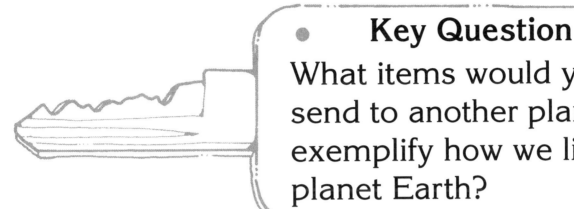

Key Question

What items would you send to another planet to exemplify how we live on planet Earth?

Learning Goal

Students will:

determine what items they would send in a time capsule to extraterrestrial life on another planet.

169

SPACE CAPSULE

What would you send to intelligent life on a planet outside our solar system?

Think about what you would send to represent human life on Earth. Write down your responses. Cut apart the strips, place the strips into groups. Tabulate and display your results by graphing.

Famous Person:	Important Invention:
Past Event:	Appliance:
Movie:	Electronic Equipment:
Television Program:	Sample Transportation:
Book :	Sample Sport Equipment :
Comic Book :	Game :
Music :	Toy :
Food :	Your Choice :

0010010101101001001010 04-0319-54

SPACE CAPSULE
SURVEY RESULTS

TOP 3 RESPONSES

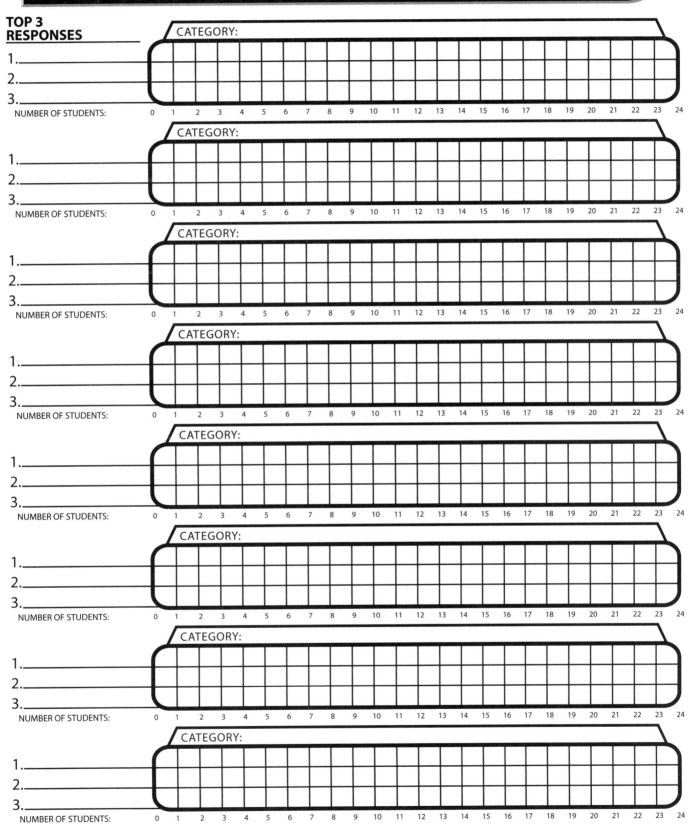

CATEGORY:

1.
2.
3.

NUMBER OF STUDENTS: 0 1 2 3 4 5 6 7 8 9 10 11 12 13 14 15 16 17 18 19 20 21 22 23 24

CATEGORY:

1.
2.
3.

NUMBER OF STUDENTS: 0 1 2 3 4 5 6 7 8 9 10 11 12 13 14 15 16 17 18 19 20 21 22 23 24

CATEGORY:

1.
2.
3.

NUMBER OF STUDENTS: 0 1 2 3 4 5 6 7 8 9 10 11 12 13 14 15 16 17 18 19 20 21 22 23 24

CATEGORY:

1.
2.
3.

NUMBER OF STUDENTS: 0 1 2 3 4 5 6 7 8 9 10 11 12 13 14 15 16 17 18 19 20 21 22 23 24

CATEGORY:

1.
2.
3.

NUMBER OF STUDENTS: 0 1 2 3 4 5 6 7 8 9 10 11 12 13 14 15 16 17 18 19 20 21 22 23 24

CATEGORY:

1.
2.
3.

NUMBER OF STUDENTS: 0 1 2 3 4 5 6 7 8 9 10 11 12 13 14 15 16 17 18 19 20 21 22 23 24

CATEGORY:

1.
2.
3.

NUMBER OF STUDENTS: 0 1 2 3 4 5 6 7 8 9 10 11 12 13 14 15 16 17 18 19 20 21 22 23 24

CATEGORY:

1.
2.
3.

NUMBER OF STUDENTS: 0 1 2 3 4 5 6 7 8 9 10 11 12 13 14 15 16 17 18 19 20 21 22 23 24

SPACE CAPSULE

04-0319-54

Connecting Learning

1. How did you decide what to put for each category?

2. How did your group come to an agreement on one response for each category?

3. Do you think this was a fair method to use? Why or why not?

4. Are you happy with the responses your group selected? Why or why not?

5. How did your group's responses compare to the other groups?

6. How do you think our classes' responses would compare to students in the same grade somewhere else in the country or the world?

7. What are you wondering now?

Topic
Rates

Key Question
How can you determine how long it will take and how much it will cost to drive around the equator of each of the planets in our solar system?

Learning Goals
Students will:
- estimate how long it will take and how much it will cost to drive around the equator of each planet in the solar system, and
- complete the necessary calculations to determine the actual cost and time it would take.

Guiding Documents
Project 2061 Benchmarks
- *Mathematical statements can be used to describe how one quantity changes when another changes. Rates of change can be computed from magnitudes and vice versa.*
- *Nine planets of very different size, composition, and surface features move around the sun in nearly circular orbits. Some planets have a great variety of moons and even flat rings of rock and ice particles orbiting around them. Some of these planets and moons show evidence of geological activity. The earth is orbited by one moon, many artificial satellites, and debris.*
- *The earth is one of several planets that orbit the sun, and the moon orbits around the earth.*

NRC Standards
- *The earth is the third planet from the sun in a system that includes the moon, the sun, eight other planets and their moons, and smaller objects, such as asteroids and comets. The sun, an average star, is the central and largest body in the solar system.*
- *Use mathematics in all aspects of scientific inquiry.*

NCTM Standards 2000*
- *Solve simple problems involving rates and derived measurements for such attributes as velocity and density*
- *Understand and use ratios and proportions to represent quantitative relationships*
- *Solve problems that arise in mathematics and in other contexts*

Math
Whole number operations
   division
    multiplication
Rates
Estimation
   rounding
Using formulas

Science
Earth science
   astronomy
    planets

Integrated Processes
Observing
Comparing and contrasting
Collecting and recording data
Interpreting data
Applying

Materials
For the class:
   calculators

Background Information

Planet	Circumference in kilometers	Diameter in kilometers	Hours
Mercury	15,390	4900	159
Venus	38,000	12,100	392
Earth	40,200	12,800	414
Mars	21,350	6800	220
Jupiter	449,020	143,000	4629
Saturn	378,680	120,600	3904
Uranus	160,450	51,100	1654
Neptune	155,430	49,500	1602
Total			12,974

Planet	Days	Liters of Fuel	Cost
Mercury	7	1026	$923.40
Venus	16	2533	$2279.70
Earth	17	2680	$2412.00
Mars	9	1423	$1280.70
Jupiter	193	29,935	$26,941.50
Saturn	163	25,245	$22,720.50
Uranus	69	10,697	$9627.30
Neptune	67	10,362	$9325.80
Total	544	84,382	$75,510.90

A Land Rover gets 15 kilometers per liter of fuel at a cost of $.90 per liter, and travels at a speed of 97 kilometers per hour.

Management
1. This activity may take 45-90 minutes to complete depending on the arrangement of groups.
2. This activity may be done individually, in small groups, or as a whole class with different students doing the calculations for assigned planets.

Procedure
1. Explain to the students that they will be traveling around the equator of each planet in a Land Rover. The surface texture of each planet will not be taken into consideration. The Land Rover will travel at a speed of 97 kilometers per hour and fuel will cost $.90 per liter. It gets 15 kilometers per liter of fuel.
2. Have the students estimate the number of days it will take to go around the equator. Also have the students estimate the amount of fuel they will need for their journey.
3. Have the students compute the cost of fuel for the trip.
4. Explain to the student how to use the different formulas to answer the questions on the student page. The students will need to round every number to the nearest whole number except the final cost.

Connecting Learning
1. How close was your estimate to what it really costs?
2. How close was your estimate to how many days it will take?
3. Around which planet's equator would you like to travel? Why? Do you think it would be a good investment? Would there be any extra costs? What would they be?
4. What are you wondering now?

Extensions
1. Have the students design their own Land Rovers. Ask them to explain why they designed them the way they did.
2. Have the students keep logs of their journeys and the adventures that happened.

*   Reprinted with permission from *Principles and Standards for School Mathematics*, 2000 by the National Council of Teachers of Mathematics. All rights reserved.

Key Question

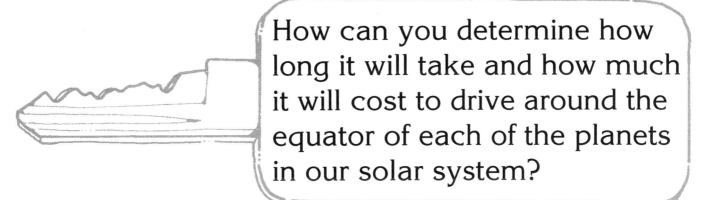

How can you determine how long it will take and how much it will cost to drive around the equator of each of the planets in our solar system?

Learning Goals

Students will:

- estimate how long it will take and how much it will cost to drive around the equator of each planet in the solar system, and
- complete the necessary calculations to determine the actual cost and time it would take.

AROUND THE PLANETS IN HOW MANY DAYS?

Imagine that it would be possible to travel in a Land Rover around the equator of each planet in our solar system. Your Land Rover can travel at a speed of 97 km per hour, gets 15 km per liter of fuel, and the fuel costs $.90 per liter. First, estimate the number of days and the cost of the fuel it would take for each planet, then compute to find the actual time and cost.

PLANET ▶	Estimation of Days	Estimation of Cost	Approximate Circumference in kilometers at Equator (C)	# Hours (C ÷ 97 km/h)	# Days (Hours ÷ 24)	Fuel Needs (C ÷ 15 = # Liters)	Fuel Costs (#Liters x $.90)
Mercury			15,390				
Venus			38,000				
Earth			40,200				
Mars			21,350				
Jupiter			449,020				
Saturn			378,680				
Uranus			160,450				
Neptune			155,430				
				TOTAL:			TOTAL:

AROUND THE PLANETS IN HOW MANY DAYS?

Connecting Learning

1. How close was your estimate to what it really costs?

2. How close was your estimate to how many days it will take?

3. Around which planet's equator would you like to travel? Why? Do you think it would be a good investment? Would there be any extra costs? What would they be?

4. What are you wondering now?

     177

ROUND AND ROUND

Topic
Geometry

Key Question
What is the shape of a planet's orbit?

Learning Goals
Students will:
- construct the shape of the planets' orbits in our solar system by drawing ellipses, and
- study the effects of several variables on the shape of the ellipses.

Guiding Documents
Project 2061 Benchmark
- *Geometric figures, number sequences, graphs, diagrams, sketches, number lines, maps, and stories can be used to represent objects, events, and processes in the real world, although such representations can never be exact in every detail.*

NCTM Standards 2000*
- *Identify, compare, and analyze attributes of two- and three-dimensional shapes and develop vocabulary to describe the attributes*
- *Build and draw geometric objects*
- *Make and test conjectures about geometric properties and relationships and develop logical arguments to justify conclusions*

Math
Geometry
  ellipses
Limits
Measurement
  length

Science
Earth science
  astronomy
    planetary orbits

Integrated Processes
Observing
Comparing and contrasting
Collecting and recording data
Interpreting data
Generalizing

Materials
For the class:
  pushpins
  string
  pencils
  cardboard sheets, 8.5" x 11"
  metric rulers

Background Information
Planetary orbits are generally elliptical. An ellipse has two foci. In this investigation, those two foci are represented by pushpins. If the foci are separated by a distance equal to one-half the length of the closed loop, then the ellipse will be at one limit: a straight line. Note that the loop would then be drawn tight. If the pushpins would have no thickness, the loop would just be a double line. The other limit is reached when the two foci are at one position. In that case, the result would be a circle.

In the first investigation, the length of string forming the loop will be 20 centimeters long and remain as the constant. The manipulated variable will be the distance between the foci, ranging from where both are at the same position (only one pin should be inserted) to where they are separated by 10 centimeters (half the distance of the length of the string), in which case a straight line results.

In the second investigation, the length of string will be the manipulated variable and the distance between the foci the constant. The shorter the string, the flatter the ellipse and the longer the string, the more circular will be the ellipse.

Management
1. It is best to do this activity in centers with three to four students in each group.
2. Students can alternate drawing the ellipses.

Procedure
1. Ask the *Key Question* and state the *Learning Goals*.
2. Supply each group with the two student pages, two pushpins, cardboard backing, string, and a metric ruler.
3. Have students tie the string very carefully to produce loops of specified sizes.
4. Demonstrate how to make the construction as shown in the illustration. Direct students to follow the instructions found on the top of the activity

pages. The pencil must be held tightly against the string to draw the line.
5. Have students complete both investigations.

Connecting Learning
1. If the length of the loop is held constant, what happens as the distance between the foci increases?
2. When does it become a line? ...a circle?
3. If the distance between the foci is held constant, what happens as the length of the loop increases?
4. What are you wondering now?

Extension
Have the students use different colored construction paper to make the different ellipses described below or others decided upon by the class. Then have them mount the smallest onto the next larger, etc., to create a colorful display. Here is an example of patterns.

Centers	Tacks-Distance	String Length
#1	9 cm	25 cm
#2	8 cm	25 cm
#3	12 cm	30 cm
#4	15 cm	35 cm
#5	13 cm	35 cm
#6	11 cm	40 cm
#7	14 cm	40 cm
#8	10 cm	45 cm
#9	12 cm	50 cm

* Reprinted with permission from *Principles and Standards for School Mathematics*, 2000 by the National Council of Teachers of Mathematics. All rights reserved.

0015

OUT OF THIS WORLD

0010010101101001001010

ROUND AND ROUND

00100101011010010010101

121-7-1985

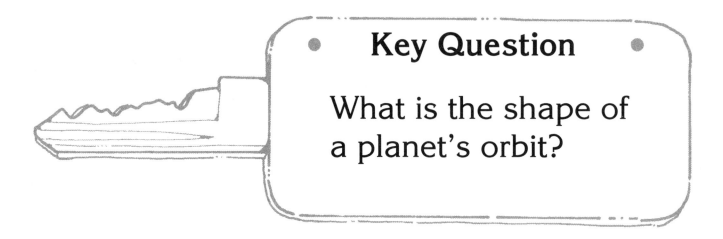

Key Question

What is the shape of a planet's orbit?

Learning Goals

Students will:

- construct the shape of planets' orbits in our solar system by drawing ellipses, and
- study the effects of several variables on the shape of the ellipses.

ROUND AND ROUND

Tie a string so that its total length is 20 cm (10 cm doubled up). Place this paper on a piece of cardboard. Push in two pins (foci), one at F and one at F'. Place your string around the two pins and draw your first ellipse. Then, successively, place your pins at EE', DD', CC', BB', and A (foci are together).

10cm

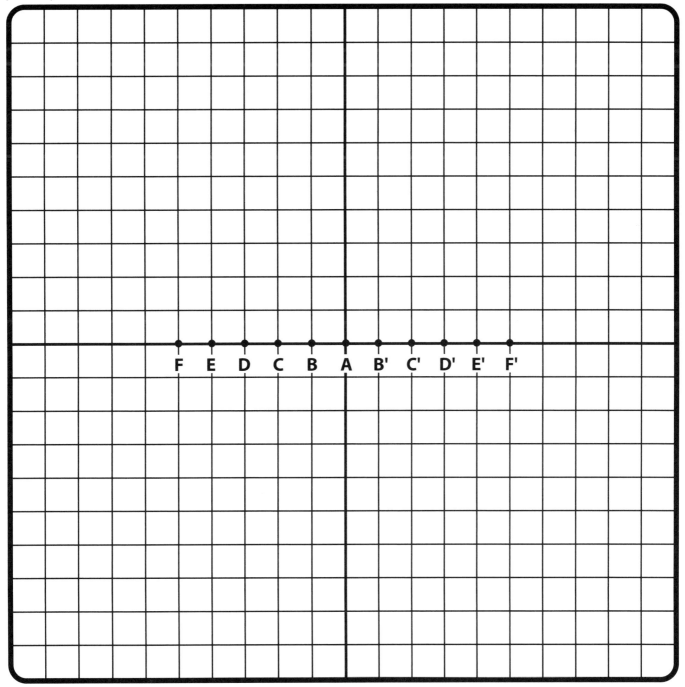

F E D C B A B' C' D' E' F'

Conclusion:

ROUND AND ROUND

In this investigation, the variable will be the length of string. Place the pins at E and E'. Draw each ellipse using these foci. Change the length of string each time. Start with 16 cm (straight line) and then test 18 cm, 20 cm, 22 cm, and 24 cm lengths.

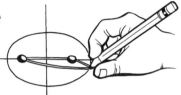

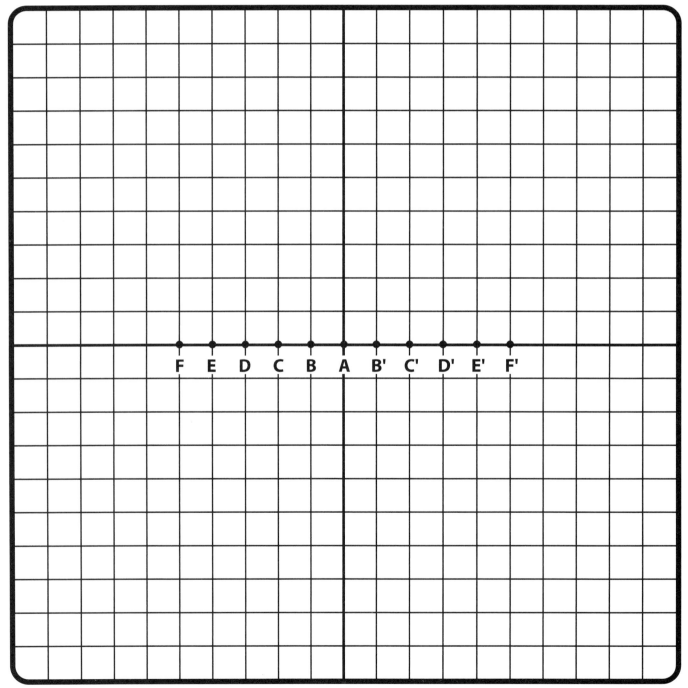

F E D C B A B' C' D' E' F'

Conclusion:

ROUND AND ROUND

121-7-1985

Connecting Learning

1. If the length of the loop is held constant, what happens as the distance between the foci increases?

2. When does it become a line? ...a circle?

3. If the distance between the foci is held constant, what happens as the length of the loop increases?

4. What are you wondering now?

0017
001001010110100100101010
STARS IN THE MILKY WAY GALAXY
OUT OF THIS WORLD
001001010110100100101010
1-00-91941

Topic
Sampling

Key Question
How do scientists estimate the number of stars in our galaxy?

Learning Goals
Students will:
- use random sampling to estimate the number of characters on a page of classified ads, and
- use proportional reasoning to determine how many pages of classified ads it would take for the number of characters to equal the number of stars in the Milky Way galaxy.

Guiding Documents
Project 2061 Benchmarks
- *Find the mean and median of a set of data.*
- *The mean, median, and mode tell different things about the middle of a data set.*
- *The larger a well-chosen sample is, the more accurately it is likely to represent the whole. but there are many ways of choosing a sample that can make it unrepresentative of the whole.*
- *Use calculators to compare amounts proportionally.*

NRC Standards
- *Different kinds of questions suggest different kinds of scientific investigations. Some investigations involve observing and describing objects, organisms, or events; some involve collecting specimens; some involve experiments; some involve seeking more information; some involve discovery of new objects and phenomena; and some involve making models.*
- *Use appropriate tools and techniques to gather, analyze, and interpret data.*
- *Mathematics is important in all aspects of scientific inquiry.*

NCTM Standards 2000*
- *Use measures of center, focusing on the median, and understand what each does and does not indicate about the data set*
- *Collect data using observations, surveys, and experiments*
- *Understand and use ratios and proportions to represent quantitative relationships*

- *Solve problems involving scale factors, using ratio and proportion*
- *Solve problems that arise in mathematics and in other contexts*

Math
Estimation
Data collection
  surveying
Data analysis
  mean
  median
  range
Proportional reasoning
  ratios
Problem solving

Science
Earth science
  astronomy
    stars

Integrated Processes
Observing
Predicting
Collecting and recording data
Interpreting data
Analyzing
Applying

Materials
For each group:
  one page of classified ads (see *Management 4*)
  meter stick or tape
  sampling squares (see *Management 2*)

For each student:
  Our Awesome Milky Way Galaxy rubber band book
  #19 rubber band
  student pages
  calculator

Background Information
There are two principal ways of gathering quantitative data—by census or by sample. In a census, every organism, object, event, etc., is counted. Since it is usually impractical or impossible to count every element, the preferred technique is sampling. For example, rather than count all the grains of sand on the

beach, you can count the number of grains in a small area. Once that value is known, you can multiply to estimate the number of grains of sand on the beach. One of the most frequently used methods of sampling is *random sampling*. In random sampling, each element has an equal chance of appearing in the sample.

Management
1. Students need to work in groups of two to six students.
2. Copy the page of sampling squares onto card stock. Each group will need six squares.
3. Six random samples is the minimum number required to get reliable data. If you choose to do more, each group will need more sampling squares.
4. Each group will need one page from the classified section of the newspaper. Try to select pages that are mostly text and don't have a lot of pictures or large ads.
5. When counting the number of characters within a square, everything counts (letters, symbols, punctuation marks, etc.). If half or more of a character falls within the square, it should be counted.

Procedure
1. Ask the *Key Question* and state the *Learning Goals*.
2. Distribute the *Our Awesome Milky Way Galaxy* rubber band book and a #19 rubber band to each student. Show students how to fold the pages into fourths and nest them together to make a book. The rubber band around the spine holds the pages together.
3. Read through the information as a class, and discuss how students think that scientists estimate the number of stars in our galaxy. Talk about the use of sampling to estimate quantities too large to count.
4. Tell students that, since they cannot count stars during the day, they are going to sample an artificial sky by counting the number of characters on a page from the classified section of the newspaper.
5. Have students get into groups and distribute the classified pages. Ask students to estimate the number of characters on the page and to begin counting.
6. After a short period of time, students should realize that there must be an "easier way" to count the characters. Ask them to suggest some possible methods.
7. Distribute the student pages, meter sticks, and the sampling squares. Have the students record their individual estimates of the number of characters on their classified pages, then average their group's estimates.
8. Have students measure the length and width of the printed portion of the page and calculate its area.

9. Instruct groups to lay their newspapers flat on the floor. Tell them to select one person to stand about 25 centimeters from the edge of the paper and toss the sampling squares onto the page. If any squares miss the paper entirely, they should be dropped again until they are on the page.
10. Have them trace around the edge of each square with a pencil and remove the squares.
11. Tell students that they now need to count the number of characters in each square. Explain that everything counts (letters, symbols, punctuation marks, etc.). If half or more of a character falls within the square, it should be counted.
12. Have students complete the student pages, using their calculators as needed to do the calculations.

Connecting Learning
1. How many "stars" were on your page?
2. Did each group come up with the same number? Why or why not?
3. How many years would you need to publish 100 billion characters in the classified ads? How does your estimate compare to the estimates of other groups?
4. What patterns do you see in the table on the final student page? Were you able to multiply all numbers by two and six respectively? Explain.
5. What year would publication have needed to begin to reach 200 billion "stars?" …600 billion?
6. Would it have been possible to publish newspapers in those years? Why or why not?
7. How could astronomers use the principle of random sampling to count the number of stars in our galaxy?
8. What are you wondering now?

Extensions
1. Use random sampling to estimate the number of grains of sand in a sandbox or the number of blades of grass on the playground.
2. Research the latest information about our galaxy and how it compares to other galaxies that have been discovered.

Internet Connections
Star Child
http://starchild.gsfc.nasa.gov/
NASA's space information website for grade-school students, with information presented at two levels.

Imagine the Universe
http://imagine.gsfc.nasa.gov/
NASA's website for ages 14 and up, or anyone wanting to learn more about the universe.

* Reprinted with permission from *Principles and Standards for School Mathematics*, 2000 by the National Council of Teachers of Mathematics. All rights reserved.

Key Question

How do scientists estimate the number of stars in our galaxy?

Learning Goals

Students will:

- use random sampling to estimate the number of characters on a page of classified ads, and

- use proportional reasoning to determine how many pages of classified ads it would take for the number of characters to equal the number of stars in the Milky Way Galaxy.

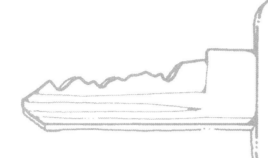

OUR AWESOME MILKY WAY GALAXY

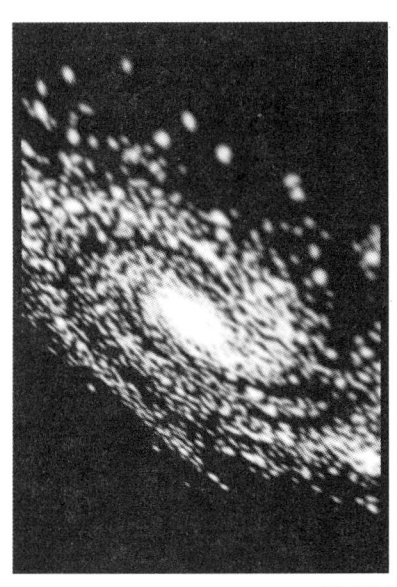

Stars in the Big Dipper

STARS ▲	Distance from Earth
MIZAR	59 LIGHT YEARS
ALCOR	59 LIGHT YEARS
MERAK	62 LIGHT YEARS
ALIOTH	62 LIGHT YEARS
MEGREZ	65 LIGHT YEARS
PHECDA	75 LIGHT YEARS
DUBHE	75 LIGHT YEARS
ALKAID	110 LIGHT YEARS

Step out during the early evening on a clear, moonless, winter night, and you will see a faint milky band of light stretching across the heavens directly overhead.

HEY! DID SOMEONE FORGET TO CLEAN THE LENS OR WHAT!?

Strangely enough, we know less about our galaxy than some of those farther out in space because we cannot view it from outside and are limited in our ability to penetrate it. Interstellar dust clouds prevent us from seeing very far into the Milky Way Galaxy even with the use of our most powerful telescopes.

(See the end of this book for distance information on the stars that make up the Big Dipper.)

Some stars are 100 times more massive than our sun. So why do these huge stars seem so small when we see them? The answer to that question is because they are so FAR away—far, far away!

The Greek scholar Democritus was the first to guess the true nature of the Milky Way. In the fourth century B.C. he wrote, "It is a lustre of small stars very close together." Over one thousand years later, Galileo turned his new telescope to look at the Milky Way to prove that Democritus had guessed right!

Galaxies come in three basic shapes: spiral, elliptical, and irregular. The Milky Way Galaxy is an open spiral galaxy with its spiral arms wrapped around a nucleus.

If we could view the Milky Way Galaxy from "above," it would look something like this drawing. Our solar system is located in one of the spiral arms.

Known as the Milky Way, this band contains the densest concentration of stars in our galaxy. The ancients imagined it was a trail of milk across the sky. All the other stars you see also belong to this gigantic star system we call the Milky Way Galaxy.

The Earth is one of the eight planets revolving around the sun, our closest star.

But our sun is only one of the more than 100 billion visible stars in the Milky Way Galaxy held together by the force of gravity!

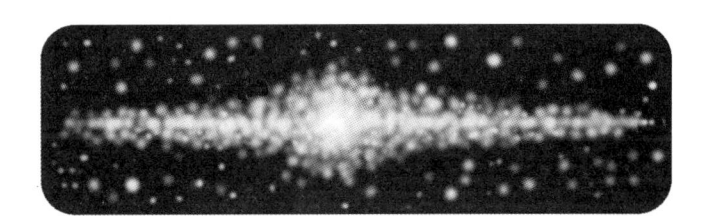

The Milky Way Galaxy is shaped like a huge disk with a spherical bulge at its center. Its diameter is estimated to be 100,000 light years*. Its thickness varies between 10,000 and 30,000 light years. Our sun is about 30,000 light years from the center. If viewed from the edge of the disk, astronomers have evidence that our galaxy looks like this.

* A light year is the distance light travels (at its speed of 186,000 miles per second) in a year.

Our galaxy is revolving around its center. It takes 200 million years to complete one revolution. Since the Earth is moving toward the edge of the revolving disk, we are speeding through space at a phenomenal rate!

Our galaxy has plenty of company. There are more than 100 billion galaxies in the universe! They typically consist of more than 100 billion visible suns with some having as many as 300 billion.

100 BILLION...
...100...

WOW, THAT'S A LOT.

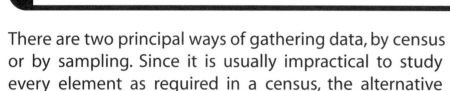

STARS IN THE MILKY WAY GALAXY

1-00-91941

Solving any problem in statistics involves three steps:

a. **definition of the problem**
b. **collection of data**
c. **analysis of the data**

There are two principal ways of gathering data, by census or by sampling. Since it is usually impractical to study every element as required in a census, the alternative technique is sampling. For example, rather than counting all of the characters on a newspaper classified ad page, one can count the number of characters in a small area and then use proportional reasoning to mathematically calculate an estimate of the total number on the page.

One of the most frequently used methods of sampling involves random samples in which each element has an equal chance of being chosen. This investigation involves random sampling.

Your Task

Examine a page from the classified section of a newspaper. Your task is to determine the approximate number of characters on the page by using sampling. Each character (a letter, symbol, or punctuation mark) counts as one. Begin by making an estimate.

I estimate that there are _____ characters on the page.

The average estimate in our group is _____ characters.

Collection of Data

1. Determine the area of the printed portion of your page.

   _____ cm x _____ cm = _____ cm²
   length   width   area

0017

OUT OF THIS WORLD
00100101011010010010101010
0010010101101001001010

STARS IN THE MILKY WAY GALAXY

00100101011010010010101010
1-00-91941

2. Cut out six squares that measure 2 cm x 2 cm and have an area of 4 square centimeters.

3. Lay the classified ad page flat on the floor. While standing about 25 centimeters from the edge of the page, toss each of the six squares onto the printed portion. All must land within the printed portion of the page to be considered. If squares fall outside the printed portion, toss them again. Carefully trace an outline around each square.

4. Count the number of characters in each square. Where characters are split by the boundary, they are counted only if half or more than half of the character lies within the square.

5. Find the average number of characters in a square.

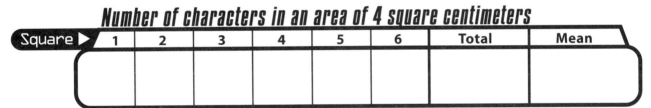

Number of characters in an area of 4 square centimeters

Square ▶	1	2	3	4	5	6	Total	Mean

Analysis of Data

1. The mean number of characters per 4 square centimeters is _____.

2. The mean number of characters per square centimeter is _____.

3. The median number of characters per 4 square centimeters is _____. (Need some help? The median value is the number in the middle of an ordered set. To find that value, arrange the numbers in increasing order and pick the middle one. Because there are six numbers, you will have to pick the middle two numbers, add them, and divide by two).

4. The median number of characters per square centimeter is _____.

5. How do the median and mean compare?

0017

OUT OF THIS WORLD

001001010110100100101010

STARS IN THE MILKY WAY GALAXY

001001010110100100101010

1-00-91941

6. The range of characters per square centimeter is_____.
   Why do you think there is such a variety of answers?_____
   _____

Calculator Fun

Use your calculator to find these answers:

7. The calculated number of characters on the page is _____.

8. What factors might have affected the outcome?

   _____

   _____

   _____

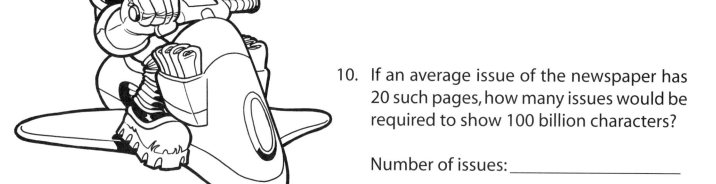

9. Using your data, how many such pages would be required to hold 100,000,000,000 characters, the estimated number of visible suns or stars in the Milky Way Galaxy, our home?

   Number of pages:_____

10. If an average issue of the newspaper has 20 such pages, how many issues would be required to show 100 billion characters?

    Number of issues: _____

0017

00100101011010010010101010

OUT OF THIS WORLD

STARS IN THE MILKY WAY GALAXY

0010010101101001001010101

1-00-91941

11. If 365 issues are published annually, how many years of publication would be required to display 100 billion characters?

    Number of years: _____

12. If such a newspaper had been published long enough to reach the 100 billion character level today, when would it have had to begin publishing?

    Number of years: _____

First Moon Landing

1969

13. Relate that year to an event in history.

    _____

    _____

    _____

14. If the longest wall in your classroom was papered with such pages, what would be the estimated number of characters on the wall?

    Number of characters on the wall: _____

15. How many such wall spaces would be required to display 100 billion characters?

    Number of walls: _____

OUT OF THIS WORLD

STARS IN THE MILKY WAY GALAXY

0017

Many researchers believe that our galaxy contains five to 10 times as much dark matter as visible stars. This mass, they believe, is equivalent to at least 600 billion stars. Other scientists believe there are closer to 200 billion visible stars in our galaxy. To try to understand these vast differences, fill in the chart to see what information we get when we compare and contrast the various viewpoints:

100,000,000,000 visible stars,
200,000,000,000 visible stars,
600,000,000,000 invisible stars in dark matter.

NUMBER OF STARS ▶	Characters per page	Number of pages required	Number of issues required	Years of publication required	Beginning year of publication
100 Billion					
200 Billion					
600 Billion					

List two things you find remarkable about your findings.

Connecting Learning

1. How many "stars" were on your page?

2. Did each group come up with the same number? Why or why not?

3. How many years would you need to publish 100 billion characters in the classified ads? How does your estimate compare to the estimates of other groups?

4. What patterns do you see in the table on the final student page? Were you able to multiply all numbers by two and six respectively? Explain.

5. What year would publication have needed to begin to reach 200 billion "stars?" ...600 billion?

0017

0010010101101001001010

OUT OF THIS WORLD

STARS IN THE MILKY WAY GALAXY

0010010101101001001010
1-00-91941

Connecting Learning

6. Would it have been possible to publish newspapers in those years? Why or why not?

7. How could astronomers use the principle of random sampling to count the number of stars in our galaxy?

8. What are you wondering now?

Topic
Constellations

Key Question
How can you construct a model of the Big Dipper to show the actual relative locations of the stars in space?

Learning Goals
Students will:
- discover that the star patterns seen in constellations are the result of the earthbound observer's perspective rather than evidence of any actual relationship between the individual stars themselves,
- construct a three-dimensional model of the Big Dipper,
- observe the model from different observation points, and
- compare results of their observations.

Guiding Documents
Project 2061 Benchmarks
- *Stars are like the sun, some being smaller and some larger, but so far away that they look like points of light.*
- *Different models can be used to represent the same thing. What kind of a model to use and how complex it should be depends on its purpose. The usefulness of a model may be limited if it is too simple or if it is needlessly complicated. Choosing a useful model is one of the instances in which intuition and creativity come into play in science, mathematics, and engineering.*
- *Geometric figures, number sequences, graphs, diagrams, sketches, number lines, maps, and stories can be used to represent objects, events, and processes in the real world, although such representations can never be exact in every detail.*
- *Scale drawings show shapes and compare locations of things very different in size.*

NCTM Standards 2000*
- *Select and apply appropriate standard units and tools to measure length, area, volume, weight, time, temperature, and the size of angles*
- *Use representations to model and interpret physical, social, and mathematical phenomena*
- *Make and use coordinate systems to specify locations and to describe paths*

Math
Measurement
Graphing
Geometry

Science
Earth science
  astronomy
    stars

Integrated Processes
Observing
Predicting
Collecting and recording data
Comparing and contrasting
Generalizing
Analyzing

Materials
For each group:
  cardboard box (i.e., copy paper box)
  7 squares of aluminum foil (3" x 3")
  scissors or craft knife
  large-eyed needle (e.g., #14 raffia needles)
  drawing compass
  graph paper
  student pages

For the class:
  transparent tape
  1 spool black upholstery thread or monofilament line

Background Information
Astronomers today recognize 88 constellations in the night skies. To an astronomer, the word *constellation* refers not only to the specific stars that make up a pattern, but the entire region of the sky in which the pattern is found. To most non-astronomers, constellations are star groups that form patterns, similar to dot-to-dot pictures, in the sky. The shape of the patterns, the stars included in the constellations, and the figures represented by the groupings are not the workings of an exact science. Rather, they are the result of an aesthetic occupation; and one's imagination must be used to see, as the early stargazers did, the heroes, kings, queens, birds, bears, bulls, and dragons on the black ceiling of the sky.

Different civilizations saw different patterns in the star groups; however, the "official" constellations in the Northern Hemisphere that are recognized today are mostly those identified by the ancient Greeks and other cultures in Europe, the Middle East, and northern Africa. These constellations include the 12 constellations of the zodiac that mark the path of the sun, moon, and planets across the heavens; Orion; Ursa Major and Ursa Minor (which include the Big Dipper and the Little Dipper); Draco, the Dragon; Cassiopeia, the Queen; and Cepheus, the King.

Because the skies of the Southern Hemisphere were not visible in the latitudes where ancient astronomers made their observations, the constellations there do not have as much history. In many cases, they were named by European explorers in the 1800s. Rather than being based upon historical or romantic stories, they are named for such scientific instruments as the telescope, the sextant, the clock, and the microscope.

Whatever the name, whatever the pattern, each constellation appears the way it does simply because of our perspective from Earth. The heavens appear to us as a great domed ceiling, much as we would view the inside of an inverted bowl were we to be beneath it. As such, each star group appears to lie in a single plane as if painted onto the ceiling of the sky.

In reality there is not a ceiling; the star groups are three-dimensional, and each group would appear completely different (many would no longer be grouped together at all) if viewed from another point in space.

Perhaps the best known of the constellations in the northern sky is the Big Dipper. In the strictest sense, the Big Dipper is not a complete constellation in itself; rather it is an *asterism*—a group of stars that forms a part of a larger constellation. In this case the constellation is Ursa Major, the Great Bear. The Big Dipper section of Ursa Major consists of seven principal stars, all fairly bright, grouped in a manner resembling the outline of a water dipper.

This star group is so conspicuous, and therefore easy to find, that it is known to most people. It is also one of the few star patterns that really look like the thing after which it was named. It is a circumpolar constellation, which means that it is in the northern sky above the horizon for some part of the night at most northern latitudes, and can be seen any night that the skies are clear and stars are visible.

Distances in space are so great that to give distances in terms of miles would be cumbersome and difficult to comprehend. Instead, astronomers talk in terms of light years. A light year is a measure of the distance light travels in one year, at a rate of 186,282 miles per second. That amounts to about 6 trillion (6,000,000,000,000) miles. For example, it is easier to say that our closest neighboring star in the Alpha Centauri system is 4.3 light years away than it is to write that it is 25,000,000,000,000 miles from Earth.

Management
1. This activity may take two class periods. Time will vary depending on students' experience working with graphs and construction of the models.

2. Students can work in small groups or in pairs if enough boxes are available.
3. Each group of students will need one copy of the scaled grid paper and four copies of the plain grid paper provided. Each student should have his or her own copy of the remaining student pages.
4. Students need to be aware that the measurements given for the constellations are approximate, as are the distances given, but that they will be able to recognize constellations if they are accurate in their graphing and measuring.
5. There are actually eight stars in the Big Dipper, but only seven of them are identified in this activity. Mizar and Alcor are both 59 light years from Earth and very close together. Mizar is the one identified on the student pages.
6. Using craft knives for cutting the holes in the cardboard boxes is easier than scissors, but safety must be ensured by giving specific instructions and a demonstration in how to properly use such tools. Also, close supervision of the cutting process is vital.

Procedure
1. Begin by comparing a map and a globe to see how different types of models are designed to show the same object in understandable terms, but using very different approaches (two-dimensional maps, three-dimensional globes). Compare and contrast the representations and discuss the advantages and disadvantages of each.
2. Pose the *Key Question* and discuss possible ways to represent the stars and their actual locations in space relative to Earth and each other.
3. Tell students that they are going to have the opportunity to make a model that shows how the stars in the Big Dipper actually exist in space and how those stars would look from somewhere besides Earth.
4. Have students get into groups and distribute a sheet of scaled grid paper to each group and the first student page.
5. Give students the coordinates for each of the stars in the Big Dipper and have groups plot the locations on the grid paper. Have them label each point with the name of the star for reference.

Star	Light years from Earth	Coordinates
Alkaid	110	(3, 10.5)
Mizar	59	(10.5, 15)
Alioth	62	(16.5, 15)
Megrez	65	(24, 15)
Phecda	75	(28.5, 9)
Merak	62	(37.5, 13.5)
Dubhe	75	(36, 21)

198

6. Instruct students to calculate the distances from the Earth to the various stars and complete the table on the first student page.
7. Distribute the materials for making the constellation model and the page of instructions to each group. Go over the instructions as a class and clarify any questions.
8. Give students time to construct the three-dimensional model of the Big Dipper according to the directions. Assist students as necessary during this time.
9. Have students look at the Big Dipper through the hole representing Earth's perspective, from the side-viewing hole, and from above the box. These other two perspectives represent views of the Big Dipper from other parts of our galaxy, outside of our own solar system.
10. Instruct students to record the appearance of the constellation from these other two viewpoints by drawing each star on the grids in the background of each view.
11. Compare the two resulting representations with the original and discuss the similarities and differences.
12. Distribute the final student page and have students determine the distances between the stars.
13. Encourage students to use their imaginations and name their "new" constellations based upon their new perspectives.

Connecting Learning

1. Why do you think the stars seem to lose their three-dimensional appearance when viewed from Earth? [There is a lack of visual reference points to give the stars their three 3-D appearance; the distances are so great that we are unable to discern the differences in distances between the different stars in a constellation.]
2. How did the perspective from Earth vary from the other two perspectives?
3. How could we construct similar models for other constellations? What types of information would we need to have in order to do so? What might be limiting factors in constructing such models? [In order to construct other models we would need to know the distances to each of the stars in the constellation and the relative positions of each of those stars. Limiting factors might include great variations in the distances between stars that preclude the use of a similar model, at least in a cardboard box.]
4. Using this scale model, where would our own sun be in relation to the view hole (Earth) as compared to where the other stars are? [The closest star in our model is 59 light years away. Our sun is about 8.5 minutes away—about 1/20,000 millimeters from the view hole.]

5. Do there appear to be any relationships between the different stars in the constellation? What reasons do you give for your answers?
6. What are you wondering now?

Extensions
1. Construct three-dimensional models of other constellations.
2. Arrange for a trip to a planetarium or arrange for a stargazing evening with a focus on identifying constellations.

Resources
Bennett, Jeffrey et. al. *The Cosmic Perspective, Third Edition*. Pearson Addison-Wesley. San Francisco. 2004.

Pasachoff, Jay M. *Stars and Planets, Fourth Edition (Peterson Field Guides)*. Houghton Mifflin. New York. 1999.

* Reprinted with permission from *Principles and Standards for School Mathematics*, 2000 by the National Council of Teachers of Mathematics. All rights reserved.

0018

0010010101101001001010 10

OUT OF THIS WORLD

IT ALL DEPENDS ON YOUR POINT OF VIEW
(IT'S A MATTER OF PERSPECTIVE) 0010010101101001001010 10

0-1201-977

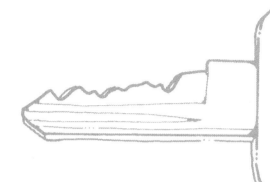

Key Question

How can you construct a model of the Big Dipper to show the actual relative locations of the stars in space?

Learning Goals

Students will:

- discover that the star patterns seen in constellations are the result of the earthbound observer's perspective rather than evidence of any actual relationship between the individual stars themselves,
- construct a three-dimensional model of the Big Dipper,
- observe the model from different observation points, and
- compare results of their observations.

Directions for Construction of the Three-Dimensional Constellation Model

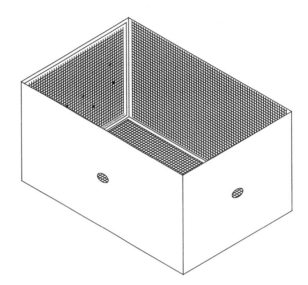

1. Tape three sheets of graph paper inside the cardboard box in the following manner:
   a. Tape the scaled grid paper with the plotted constellation in one end of the box.
   b. Tape two of the plain grid pages in the bottom of the box.
   c. Tape two plain grid pages on one side of the box.
2. Cut a hole in the end of the box opposite the plotted grid paper. Center the hole in the side of the box and make it one centimeter in diameter.
3. Using the point of a drawing compass, *carefully* poke a hole through the center of each point plotted on the graph paper in the end of the box. Poke the holes all the way through the cardboard.
4. Cut a piece of thread at least 25 centimeters longer than the length of the box and thread it through the eye of a needle.
5. Pull the needle and thread from the outside of the box through the viewing hole. Tape about 10 centimeters of the end of the thread to the outside of the box next to the hole.
6. Crumple one of the 3" x 3" squares of aluminum foil into a ball. Make the ball fairly compact, but not as tight as you can get it yet.
7. Push the needle and thread through the ball and squeeze the foil more tightly.
8. Push the needle and thread through the star hole that is closest to the center of the other side of the box (either Alioth or Megrez). Remove the needle from the thread.
9. Pull on the thread so that it is as tight as possible and tape it down to the outside of the box.
10. Measure the distance (in centimeters) from the viewing hole (Earth) to the ball (star). Slide the ball of foil along the thread until it is the distance from the viewing hole (Earth) that you calculated for the scale model. Squeeze the foil tightly to prevent it from moving.
11. Repeat steps 4–10 for each of the remaining stars, putting each star on its own thread through its own hole. Don't cross threads over each other—it makes the constellation pattern harder to see.
12. After all stars have been put in place, cut a piece of paper large enough to cover the hole in the end of the box. Punch a hole in this paper using a hole punch. Tape this piece of paper over the viewing hole so that you look through a smaller opening. This increases the sharpness of your view. (Try looking before and after you cover the hole and you'll see a difference.)
13. Cut a second hole, about one centimeter in diameter on the side of the box opposite the blank graph paper. This hole represents a view of the constellation from somewhere outside of our solar system. Leave this hole large, because if you make it smaller you won't be able to see all of the stars in the constellation.
14. You are now ready to use the model. Enjoy!

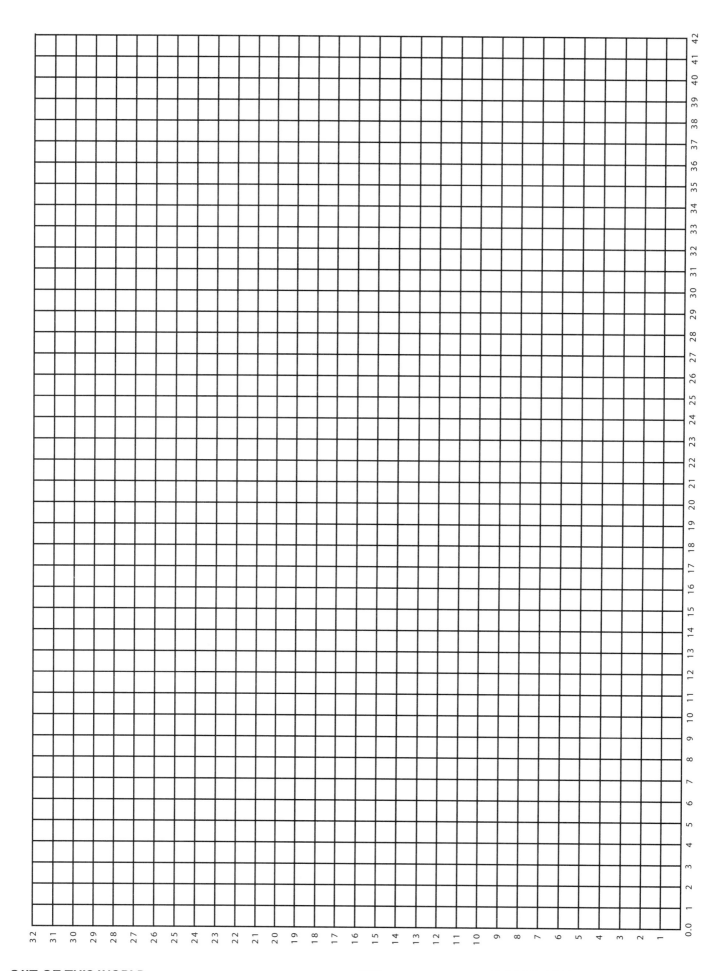

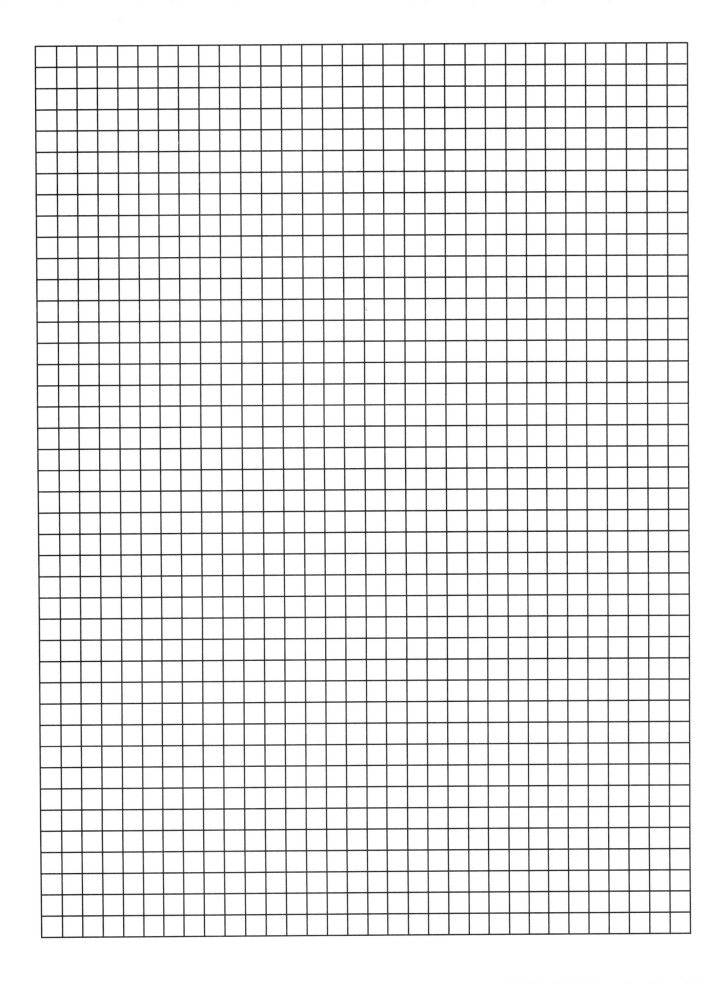

IT ALL DEPENDS ON YOUR POINT OF VIEW

DISTANCE TO THE STARS

this activity we will use a scale of 1 light year* = 3 millimeters. Creating a model using this scale lets us see the three-dimensional characteristics of the constellation and also determine the stars' distances from one another.

To determine how far a star is from Earth in our model, multiply the number of light years by 3 millimeters. This is the value you will use when constructing your model.

To determine the actual distance of the stars from Earth in miles, multiply the number of light years by 6 trillion miles.

In order to create a model of the Big Dipper that accurately shows the distance of each star from the Earth and from the other stars, the model must be made to scale. For

STAR ▶	Distance From Earth in Light Years		3 (1 Light Year = 3 mm)		Scale Model Distance From Earth in mm	Distance From Earth in Light Years		6 trillion 6,000,000,000,000		Distance From Earth in Miles
Alkaid	110	×	3	=		110	×	6 Trillion	=	
Mizar	59	×	3	=		59	×	6 Trillion	=	
Alioth	62	×	3	=		62	×	6 Trillion	=	
Megrez	65	×	3	=		65	×	6 Trillion	=	
Phecda	75	×	3	=		75	×	6 Trillion	=	
Merak	62	×	3	=		62	×	6 Trillion	=	
Dubhe	75	×	3	=		75	×	6 Trillion	=	

* A light year is a measure of the distance that light travels in one year at a rate of 186,282 miles per second. That distance is approximately 6 trillion miles (6,000,000,000,000 miles).

Using your model, you can closely determine the distances between the stars themselves. To do this, measure the distance between two stars, then convert that distance into light years by dividing by 3. For example, if two stars are 100 mm apart, divide 100 mm by 3 to get 33.3 light years. This tells you that the stars are actually 33.3 light years apart.

Calculate the distance between the following pairs of stars:

STAR PAIRS ▼	Distance in mm	÷ 3	# of Light Years Apart
Megrez-Dubhe			
Alkaid-Mizar			
Phecda-Merak			
Megrez-Alioth			

Use your model and compare measurements to answer these questions.

1. Which two stars in the constellation are the closest neighbors?

2. Which distance is greater, the distance between Dubhe and Alioth or the distance between Mizar and Phecda? How much greater?

3. Would it take longer to travel from Alkaid to Mizar or from Alkaid to Phecda? If you were traveling at the speed of light, how much longer would it take?

4. How many light years would it take you to travel from Merak over to Alioth, then to Dubhe, and finally back to Merak?

5. Both Phecda and Dubhe are 75 light years from Earth. How far apart are they from one another?

6. Which two stars are 17 light years apart?

7. Which two stars are 58 light years apart?

IT ALL DEPENDS ON YOUR POINT OF VIEW

(IT'S A MATTER OF PERSPECTIVE) 00100101011010010010101010

0-1201-977

Connecting Learning

1. Why do you think the stars seem to lose their three-dimensional appearance when viewed from Earth?

2. How did the perspective from Earth vary from the other two perspectives?

3. How could we construct similar models for other constellations? What types of information would we need to have in order to do so? What might be limiting factors in constructing such models?

4. Using this scale model, where would our own sun be in relation to the view hole (Earth) as compared to where the other stars are?

5. Do there appear to be any relationships between the different stars in the constellation? What reasons do you give for your answers?

6. What are you wondering now?

Societies/Organizations

Astronomical Society of the Pacific
http://www.astrosociety.org
390 Ashton Avenue
San Francisco, CA 94122
415-337-1100

MIRA (Monterey Institute for Research in Astronomy)
http://www.mira.org
200 Eighth Street
Marina, CA 93933
831-883-1000
mira@mira.org

Pacific Science Center
http://www.pacsci.org
200 Second Avenue North
Seattle, WA 98109
206-443-2001

NASA (National Aeronautics and Space Administration)
http://www.nasa.gov

Young Astronaut Council
http://www.youngastronauts.org/yac/
5200 27th Street NW
Washington, D.C. 20015
301-617-0923
youngastronauts@aol.com

Books

Asimov, Isaac. *Isaac Asimov's Guide to the Earth and Space.* Fawcett. New York. 1992.

Beatty, J. Kelly, Carolyn Collins Petersen, and Andrew Chaikin (Eds.). *The New Solar System, Fourth Edition.* Cambridge University Press. NY. 1998.

Bennett, Jeffrey et. al. *The Cosmic Perspective, Third Edition.* Pearson Education, Inc. San Francisco. 2004.

Cole, Joanna. *The Magic School Bus Lost in the Solar System.* Scholastic Press. New York. 1992.

Curtis, Anthony. *Space Almanac, Second Edition.* Gulf Publishing Co. Houston. 2001.

Freedman, Roger A. and William J. Kaufmann III. *Universe, Eighth Edition.* W.H. Freeman & Co. New York. 2007.

Littmann, Mark. *Planets Beyond: Discovering the Outer Solar System.* Dover Publications. New York. 2004.

Miller, Ron and William K. Hartmann. *The Grand Tour: A Traveler's Guide to the Solar System, Revised Edition.* Workman Publishing. New York. 2005.

Ride, Sally and Tam O'Shaughnessy. *Exploring Our Solar System.* Crown Books. New York. 2003.

Software

SkyGazer
Carina Software
http://www.carinasoft.com

RedShift
Maris Technologies
http://www.redshift.de

Stellarium
http://www.stellarium.org

The AIMS Program

AIMS is the acronym for "Activities Integrating Mathematics and Science." Such integration enriches learning and makes it meaningful and holistic. AIMS began as a project of Fresno Pacific University to integrate the study of mathematics and science in grades K-9, but has since expanded to include language arts, social studies, and other disciplines.

AIMS is a continuing program of the non-profit AIMS Education Foundation. It had its inception in a National Science Foundation funded program whose purpose was to explore the effectiveness of integrating mathematics and science. The project directors in cooperation with 80 elementary classroom teachers devoted two years to a thorough field-testing of the results and implications of integration.

The approach met with such positive results that the decision was made to launch a program to create instructional materials incorporating this concept. Despite the fact that thoughtful educators have long recommended an integrative approach, very little appropriate material was available in 1981 when the project began. A series of writing projects have ensued, and today the AIMS Education Foundation is committed to continue the creation of new integrated activities on a permanent basis.

The AIMS program is funded through the sale of books, products, and staff development workshops and through proceeds from the Foundation's endowment. All net income from program and products flows into a trust fund administered by the AIMS Education Foundation. Use of these funds is restricted to support of research, development, and publication of new materials. Writers donate all their rights to the Foundation to support its on-going program. No royalties are paid to the writers.

The rationale for integration lies in the fact that science, mathematics, language arts, social studies, etc., are integrally interwoven in the real world from which it follows that they should be similarly treated in the classroom where we are preparing students to live in that world. Teachers who use the AIMS program give enthusiastic endorsement to the effectiveness of this approach.

Science encompasses the art of questioning, investigating, hypothesizing, discovering, and communicating. Mathematics is the language that provides clarity, objectivity, and understanding. The language arts provide us powerful tools of communication. Many of the major contemporary societal issues stem from advancements in science and must be studied in the context of the social sciences. Therefore, it is timely that all of us take seriously a more holistic mode of educating our students. This goal motivates all who are associated with the AIMS Program. We invite you to join us in this effort.

Meaningful integration of knowledge is a major recommendation coming from the nation's professional science and mathematics associations. The American Association for the Advancement of Science in *Science for All Americans* strongly recommends the integration of mathematics, science, and technology. The National Council of Teachers of Mathematics places strong emphasis on applications of mathematics such as are found in science investigations. AIMS is fully aligned with these recommendations.

Extensive field testing of AIMS investigations confirms these beneficial results:

1. Mathematics becomes more meaningful, hence more useful, when it is applied to situations that interest students.
2. The extent to which science is studied and understood is increased, with a significant economy of time, when mathematics and science are integrated.
3. There is improved quality of learning and retention, supporting the thesis that learning which is meaningful and relevant is more effective.
4. Motivation and involvement are increased dramatically as students investigate real-world situations and participate actively in the process.

We invite you to become part of this classroom teacher movement by using an integrated approach to learning and sharing any suggestions you may have. The AIMS Program welcomes you!

AIMS Education Foundation Programs

Practical proven strategies to improve student achievement

When you host an AIMS workshop for elementary and middle school educators, you will know your teachers are receiving effective usable training they can apply in their classrooms immediately.

Designed for teachers—AIMS Workshops:

- Correlate to your state standards;
- Address key topic areas, including math content, science content, problem solving, and process skills;
- Teach you how to use AIMS' effective hands-on approach;
- Provide practice of activity-based teaching;
- Address classroom management issues, higher-order thinking skills, and materials;
- Give you AIMS resources; and
- Offer college (graduate-level) credits for many courses.

Aligned to district and administrator needs—AIMS workshops offer:

- Flexible scheduling and grade span options;
- Custom (one-, two-, or three-day) workshops to meet specific schedule, topic and grade-span needs;
- Pre-packaged one-day workshops on most major topics—only $3,900 for up to 30 participants (includes all materials and expenses);
- Prepackaged *week-long* workshops (four- or five-day formats) for in-depth math and science training—only $12,300 for up to 30 participants (includes all materials and expenses);
- Sustained staff development, by scheduling workshops throughout the school year and including follow-up and assessment;
- Eligibility for funding under the Eisenhower Act and No Child Left Behind; and

- Affordable professional development—save when you schedule consecutive-day workshops.

University Credit—Correspondence Courses

AIMS offers correspondence courses through a partnership with Fresno Pacific University.

- Convenient distance-learning courses—you study at your own pace and schedule. No computer or Internet access required!

The tuition for each three-semester unit graduate-level course is $264 plus a materials fee.

The AIMS Instructional Leadership Program

This is an AIMS staff-development program seeking to prepare facilitators for leadership roles in science/math education in their home districts or regions. Upon successful completion of the program, trained facilitators become members of the AIMS Instructional Leadership Network, qualified to conduct AIMS workshops, teach AIMS in-service courses for college credit, and serve as AIMS consultants. Intensive training is provided in mathematics, science, process and thinking skills, workshop management, and other relevant topics.

Introducing AIMS Science Core Curriculum

Developed in alignment with your state standards, AIMS' Science Core Curriculum gives students the opportunity to build content knowledge, thinking skills, and fundamental science processes.

- *Each* grade specific module has been developed to extend the AIMS approach to full-year science programs.
- *Each* standards-based module includes math, reading, hands-on investigations, and assessments.

Like all AIMS resources these core modules are able to serve students at all stages of readiness, making these a great value across the grades served in your school.

For current information regarding the programs described above, please complete the following:

Information Request

Please send current information on the items checked:

____ *Basic Information Packet* on AIMS materials   ____ Hosting information for AIMS workshops
____ *AIMS Instructional Leadership Program*   ____ AIMS Science Core Curriculum

Name _____ Phone _____

Address_____
       Street   City   State   Zip

Magazine

YOUR K-9 MATH AND SCIENCE
CLASSROOM ACTIVITIES RESOURCE

The AIMS Magazine is your source for standards-based, hands-on math and science investigations. Each issue is filled with teacher-friendly, ready-to-use activities that engage students in meaningful learning.

• *Four issues each year (fall, winter, spring, and summer).*

Current issue is shipped with all past issues within that volume.

| 1822 | Volume | XXII | 2007-2008 | $19.95 |
| 1823 | Volume | XXIII | 2008-2009 | $19.95 |

Two-Volume Combination

| M20507 | Volumes | XXI & XXII | 2006-2008 | $34.95 |
| M20608 | Volumes | XXII & XXIII | 2007-2009 | $34.95 |

Back Volumes Available
Complete volumes available for purchase:

1802	Volume II	1987-1988	$19.95
1804	Volume IV	1989-1990	$19.95
1805	Volume V	1990-1991	$19.95
1807	Volume VII	1992-1993	$19.95
1808	Volume VIII	1993-1994	$19.95
1809	Volume IX	1994-1995	$19.95
1810	Volume X	1995-1996	$19.95
1811	Volume XI	1996-1997	$19.95
1812	Volume XII	1997-1998	$19.95
1813	Volume XIII	1998-1999	$19.95
1814	Volume XIV	1999-2000	$19.95
1815	Volume XV	2000-2001	$19.95
1816	Volume XVI	2001-2002	$19.95
1817	Volume XVII	2002-2003	$19.95
1818	Volume XVIII	2003-2004	$19.95
1819	Volume XIX	2004-2005	$19.95
1820	Volume XX	2005-2006	$19.95
1821	Volume XXI	2006-2007	$19.95
1822	Volume XXII	2007-2008	$19.95

Volumes II to XIX include 10 issues.

Call 1.888.733.2467 or go to www.aimsedu.org

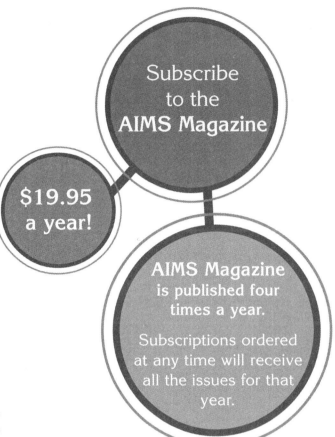

Subscribe to the AIMS Magazine

$19.95 a year!

AIMS Magazine is published four times a year.

Subscriptions ordered at any time will receive all the issues for that year.

AIMS Online—www.aimsedu.org

To see all that AIMS has to offer, check us out on the Internet at www.aimsedu.org. At our website you can search our activities database; preview and purchase individual AIMS activities; learn about core curriculum, college courses, and workshops; buy manipulatives and other classroom resources; and download free resources including articles, puzzles, and sample AIMS activities.

AIMS News
While visiting the AIMS website, sign up for AIMS News, our FREE e-mail newsletter. You'll get the latest information on what's new at AIMS including:

• New publications;
• New core curriculum modules; and
• New materials.

Sign up today!

AIMS Program Publications

Actions with Fractions, 4-9
Awesome Addition and Super Subtraction, 2-3
Bats Incredible! 2-4
Brick Layers II, 4-9
Chemistry Matters, 4-7
Counting on Coins, K-2
Cycles of Knowing and Growing, 1-3
Crazy about Cotton, 3-7
Critters, 2-5
Earth Book, 6-9
Electrical Connections, 4-9
Exploring Environments, K-6
Fabulous Fractions, 4-6
Fall into Math and Science, K-1
Field Detectives, 3-6
Finding Your Bearings, 4-9
Floaters and Sinkers, 5-9
From Head to Toe, 5-9
Fun with Foods, 5-9
Glide into Winter with Math and Science, K-1
Gravity Rules! 5-12
Hardhatting in a Geo-World, 3-5
It's About Time, K-2
It Must Be A Bird, Pre-K-2
Jaw Breakers and Heart Thumpers, 3-5
Looking at Geometry, 6-9
Looking at Lines, 6-9
Machine Shop, 5-9
Magnificent Microworld Adventures, 5-9
Marvelous Multiplication and Dazzling Division, 4-5
Math + Science, A Solution, 5-9
Mostly Magnets, 2-8
Movie Math Mania, 6-9
Multiplication the Algebra Way, 6-8
Off the Wall Science, 3-9
Out of This World, 4-8
Paper Square Geometry:
    The Mathematics of Origami, 5-12
Puzzle Play, 4-8
Pieces and Patterns, 5-9
Popping With Power, 3-5
Positive vs. Negative, 6-9
Primarily Bears, K-6
Primarily Earth, K-3

Primarily Physics, K-3
Primarily Plants, K-3
Problem Solving: Just for the Fun of It! 4-9
Problem Solving: Just for the Fun of It! Book Two, 4-9
Proportional Reasoning, 6-9
Ray's Reflections, 4-8
Sensational Springtime, K-2
Sense-Able Science, K-1
Soap Films and Bubbles, 4-9
Solve It! K-1: Problem-Solving Strategies, K-1
Solve It! 2nd: Problem-Solving Strategies, 2
Solve It! 3rd: Problem-Solving Strategies, 3
Solve It! 4th: Problem-Solving Strategies, 4
Solve It! 5th: Problem-Solving Strategies, 5
Solving Equations: A Conceptual Approach, 6-9
Spatial Visualization, 4-9
Spills and Ripples, 5-12
Spring into Math and Science, K-1
The Amazing Circle, 4-9
The Budding Botanist, 3-6
The Sky's the Limit, 5-9
Through the Eyes of the Explorers, 5-9
Under Construction, K-2
Water Precious Water, 2-6
Weather Sense: Temperature, Air Pressure, and Wind, 4-5
Weather Sense: Moisture, 4-5
Winter Wonders, K-2

Spanish Supplements*
Fall Into Math and Science, K-1
Glide Into Winter with Math and Science, K-1
Mostly Magnets, 2-8
Pieces and Patterns, 5-9
Primarily Bears, K-6
Primarily Physics, K-3
Sense-Able Science, K-1
Spring Into Math and Science, K-1

* Spanish supplements are only available as downloads from the AIMS website. The supplements contain only the student pages in Spanish; you will need the English version of the book for the teacher's text.

Spanish Edition
Constructores II: Ingeniería Creativa Con Construcciones
    LEGO® 4-9
    The entire book is written in Spanish. English pages not included.

Other Publications
Historical Connections in Mathematics, Vol. I, 5-9
Historical Connections in Mathematics, Vol. II, 5-9
Historical Connections in Mathematics, Vol. III, 5-9
Mathematicians are People, Too
Mathematicians are People, Too, Vol. II
What's Next, Volume 1, 4-12
What's Next, Volume 2, 4-12
What's Next, Volume 3, 4-12

For further information write to:
AIMS Education Foundation • P.O. Box 8120 • Fresno, California 93747-8120
www.aimsedu.org • 559.255.6396 (fax) • 888.733.2467 (toll free)

© 2007 AIMS Education Foundation

Duplication Rights

Standard Duplication Rights

Purchasers of AIMS activities (individually or in books and magazines) may make up to 200 copies of any portion of the purchased activities, provided these copies will be used for educational purposes and only at one school site.

Workshop or conference presenters may make one copy of a purchased activity for each participant, with a limit of five activities per workshop or conference session.

Standard duplication rights apply to activities received at workshops, free sample activities provided by AIMS, and activities received by conference participants.

All copies must bear the AIMS Education Foundation copyright information.

Unlimited Duplication Rights

To ensure compliance with copyright regulations, AIMS users may upgrade from standard to unlimited duplication rights. Such rights permit unlimited duplication of purchased activities (including revisions) for use at a given school site.

Activities received at workshops are eligible for upgrade from standard to unlimited duplication rights.

Free sample activities and activities received as a conference participant are not eligible for upgrade from standard to unlimited duplication rights.

Upgrade Fees

The fees for upgrading from standard to unlimited duplication rights are:
* $5 per activity per site,
* $25 per book per site, and
* $10 per magazine issue per site.

The cost of upgrading is shown in the following examples:
* activity: 5 activities x 5 sites x $5 = $125
* book: 10 books x 5 sites x $25 = $1250
* magazine issue: 1 issue x 5 sites x $10 = $50

Purchasing Unlimited Duplication Rights

To purchase unlimited duplication rights, please provide us the following:
1. The name of the individual responsible for coordinating the purchase of duplication rights.
2. The title of each book, activity, and magazine issue to be covered.
3. The number of school sites and name of each site for which rights are being purchased.
4. Payment (check, purchase order, credit card).

Requested duplication rights are automatically authorized with payment. The individual responsible for coordinating the purchase of duplication rights will be sent a certificate verifying the purchase.

Internet Use

Permission to make AIMS activities available on the Internet is determined on a case-by-case basis.

* P. O. Box 8120, Fresno, CA 93747-8120 •
* aimsed@aimsedu.org • www.aimsedu.org •
* 559.255.6396 (fax) • 888.733.2467 (toll free) •